Environmental Footprints and Eco-design of Products and Processes

Series Editor

Subramanian Senthilkannan Muthu, Green Story Inc., Toronto, Canada

Indexed by Scopus

This series aims to broadly cover all the aspects related to environmental assessment of products, development of environmental and ecological indicators and eco-design of various products and processes. Below are the areas fall under the aims and scope of this series, but not limited to: Environmental Life Cycle Assessment; Social Life Cycle Assessment; Organizational and Product Carbon Footprints; Ecological, Energy and Water Footprints; Life cycle costing; Environmental and sustainable indicators; Environmental impact assessment methods and tools; Eco-design (sustainable design) aspects and tools; Biodegradation studies; Recycling; Solid waste management; Environmental and social audits; Green Purchasing and tools; Product environmental footprints; Environmental management standards and regulations; Eco-labels; Green Claims and green washing; Assessment of sustainability aspects.

Subramanian Senthilkannan Muthu
Editor

Sustainability and Water Footprint

Industry-specific Assessments and Recommendations

Springer

Editor
Subramanian Senthilkannan Muthu
Chief Sustainability Officer
Green Story Inc.
Toronto, ON, Canada

ISSN 2345-7651 ISSN 2345-766X (electronic)
Environmental Footprints and Eco-design of Products and Processes
ISBN 978-3-031-70809-1 ISBN 978-3-031-70810-7 (eBook)
https://doi.org/10.1007/978-3-031-70810-7

This Springer imprint is published by the registered company Springer Nature Switzerland AG
The registered company address is: Gewerbestrasse 11, 6330 Cham, Switzerland

This book is dedicated to:
The lotus feet of my beloved Lord
Pazhaniandavar
My beloved late Father
My beloved Mother
My beloved Wife Karpagam and Daughters—
Anu and Karthika
My beloved Brother
Last but not least
To everyone working in various industrial
spaces to make the world "SUSTAINABLE"

Contents

Introduction to the Book—Sustainability and Water Footprint: Industry-Specific Assessments and Recommendations 1
Subramanian Senthilkannan Muthu

Industrial Grey Water Footprint: Principles, Evaluation Method, and Challenges 7
Keivan Arastou, Ali Dehnavi, and Shervin Jamshidi

Assessing the Sustainability of Agricultural Water Use Based on Water Footprints of Wheat and Rice Production 57
Shahla Dehghanpir, Ommolbanin Bazrafshan, Sina Nadi, and Sajad Jamshidi

Water Footprint (WFP) in Hydropower Plants Using Observed Meteorological Datasets: The Case of the Funil Hydropower Plant, Southeast Brazil 83
Eduardo de O. Bueno, Carlos R. Mello, Jorge A. Guzman, and Geovane J. Alves

Evaluation of the Water Footprint of the Use and Maintenance of Buildings. Application to Dwellings in Andalusia, Spain 117
Daniel R. Vaquedano-Ramírez, Patricia González-Vallejo, M. Rocio Ruiz-Perez, and Madelyn Marrero

Sustainable Brine Management and Carbon Dioxide Capture: Minimizing Water Footprint, Addressing Environmental Challenges, and Promoting Resource Recovery 145
Jawad Mustafa, Muhammad Ahmad, Muhammad Imran, and Muhammad Wakil Shahzad

Index 185

Introduction to the Book—Sustainability and Water Footprint: Industry-Specific Assessments and Recommendations

Subramanian Senthilkannan Muthu

Water is a critical resource for sustaining life, economic activities, and ecological balance. As the global population grows and urbanization and industrialization expand, the demand for freshwater resources increases, leading to water scarcity and environmental challenges. In this context, the concept of water footprint emerges as a vital tool to assess and manage water use across different sectors. The water footprint measures the amount of water used directly and indirectly by consumers and producers, including blue water (surface and groundwater), green water (rainwater), and grey water (water required to assimilate pollutants). This comprehensive metric helps in understanding the total water consumption and its impacts, thereby guiding more sustainable water management practices. This book is all about water footprint metric and how it is applied in various industrial sectors.

This book, "Sustainability and Water Footprint: Industry-Specific Assessments and Recommendations," aims to provide a detailed understanding of the water footprint, its evaluation methods, and strategies for reducing it in various industries. The chapters in this book are designed to equip readers with a thorough understanding of the principles, methodologies, challenges, and innovat`ive solutions related to water footprint management.

The second chapter will set the stage by enumerating industrial grey water footprint in terms of the principles, evaluation method, and challenges. The industrial water footprint (IWF) accounts for the two factors, the quantity and quality, through blue water footprint (BWF) and grey water footprint (GWF), respectively. In addition to the principles and methods, the second chapter explains the calculated IWF of an industrial park located in Isfahan province, Iran. It covers an area of 3.09 km^2 and includes 554 units: iron and metal-working, textile and dyeing, chemical and fertilizer production, food and dairy production, cellulose industry and pulp and paper, electrical and electronic, administration and other industrial

S. S. Muthu (✉)
Green Story Inc., Toronto, Canada

S. S. Muthu (ed.), *Sustainability and Water Footprint*, Environmental Footprints and Eco-design of Products and Processes, https://doi.org/10.1007/978-3-031-70810-7_1

units, and non-metallic minerals. For its GWF assessment, multiple pollutants were evaluated over a period of 6 months (October 2022 to March 2023). Regarding the production yields of the Industrial Park, the average water footprint of production (WFP) is 1.76 m^3/ton, derived from a net BWF of 1668 ± 54 m^3/day and a GWF of 2.155 ± 0.285 million cubic meters (MCM)/month, attributed to NH_4, as reported in this study. Based on the analyzed conventional pollutants, the dilution factor indicator (Df) measured how many times polluted water needs to be mixed with fresh water to achieve a safe concentration level. In this study, the Df for NH_4 is typically around 50 times. It has been argued that the industrial WF faces various challenges in standards, experimentation, and accounting. Furthermore, mitigating GWF and WF requires a more holistic approach, incorporating technological innovations, recycling initiatives, and sustainable practices. Although these challenges and future trends, as well as the methods applicable to the industrial park, are discussed in this chapter, they are also calculable for various and specific individual industries, as concluded by the authors of this chapter.

It is well known that, agriculture is very much linked to water footprint, and this link is dealt with in detail in the next chapter. Water footprint in the agricultural sector and its use in the evaluation of water stress indices are dealt with in the third chapter. Water footprint is widely used as an indicator for evaluating freshwater resources and managing water use in agriculture. This chapter focuses on Iran, an arid and semi-arid country facing severe water shortages, and evaluates the water footprint of wheat and rice production. It introduces water stress and water poverty indices, discussing how these indicators help assess water scarcity and manage water resources. This research was undertaken to assess water scarcity in the production of wheat and rice within Iran, focusing on the water footprint framework from 2008 to 2019. The findings indicate that in irrigated wheat production, the proportions of blue, green, and GWFs constitute 71.76%, 15.87%, and 13.11% of the overall water footprint, respectively. Conversely, for rainfed wheat production in Iran, green and GWFs represent 55% and 45%, respectively. The overall water footprint for rice amounts to 3037 m^3/ton, with blue, green, and GWFs comprising 91.68%, 6.93%, and 1.39%, respectively, as reported by the authors of this chapter. The available water resources for rice production across the provinces amount to 21,992 m^3/ton, with 15,210 and 6872 million m^3 (MCM) pertaining to green and blue water, respectively. Additionally, the study reveals that the average water stress index for wheat and rice production stands at 0.6 and 0.5, respectively. North Khorasan, Khuzestan, and Fars provinces exhibit the highest water self-sufficiency indices (indicating the lowest dependency) for wheat production, whereas Guilan and Mazandaran provinces display the lowest indices (indicating the highest dependency). In contrast, the highest (and lowest) self-sufficiency index for rice production is observed in Guilan province (and Khorasan Razavi, respectively). The estimated annual water deprivation for wheat and rice production in Iran is 1004.44 MCM and 1073 MCM, respectively. This research was undertaken to assess water scarcity in the production of wheat and rice within Iran, focusing on the water footprint framework from 2008 to 2019. The findings indicate that in irrigated wheat production, the proportions of blue, green, and GWFs constitute 71.76%, 15.87%, and 13.11% of the overall water footprint, respectively. Conversely, for rainfed wheat production in Iran, green and GWFs represent 55% and 45%, respectively. The overall water footprint for rice amounts to 3037 m^3/ton, with blue, green, and GWFs comprising 91.68%, 6.93%, and 1.39%, respectively. The available water resources for rice production across the provinces amount to 21,992 m^3/ton, with 15,210 and 6872 million m^3 (MCM) pertaining to green and blue water, respectively. Additionally, this study reveals that the average water stress index for wheat and rice production stands at 0.6 and 0.5, respectively. North Khorasan, Khuzestan, and Fars provinces exhibit the highest water self-sufficiency indices (indicating the lowest dependency) for wheat production, whereas Guilan and Mazandaran provinces display the lowest indices (indicating the highest dependency). In contrast, the highest (and lowest) self-sufficiency index for rice production is observed in Guilan province (and Khorasan Razavi, respectively). The estimated annual water deprivation for wheat and rice production in Iran is 1004.44 MCM and 1073 MCM, respectively. This study offers valuable insights for decision-making regarding the sustainable management of water resources in grain production. The findings of this study highlight a significant reliance on BWFs over green water footprints in the production of wheat and rice in Iran, underscoring the pivotal role of blue water resources in the agricultural process. This distinction illustrates a substantial consumption of surface and underground water resources, as opposed to reliance on green water resources, such as rainwater. The comparatively low green water footprint indicates that rainfall in Iran is insufficient to support the cultivation of wheat and rice, emphasizing the critical importance of water management in the agricultural sector. This scenario underscores the considerable impact that wheat and rice production have on Iran's water resources, affirming that the implications of these agricultural activities on water availability and sustainability cannot be overlooked. The study's findings suggest that the variations in the water stress index, water poverty, and water footprint, which are crucial for assessing water scarcity in wheat and rice production, are predominantly influenced by the utilization of blue water resources over green water resources.

The fourth chapter revolves around an interesting case study on the evaluation of the water footprint in a hydropower plant. In this chapter, water footprint (WFP) has been used to indicate the efficiency and sustainability of water resources in the hydropower generation system. It was developed to indicate how much water is spent to produce one unit of a given product or service. In the case of electricity, the product is the energy produced in a period (monthly or annually). However, it is unclear how much water is consumed for electric energy production in hydropower plants because this process has been taken as non-consumptive, i.e., the water used for energy production returns to the river. The major issue in the case of hydropower plants is the water stored in the reservoirs, which is exposed to solar radiation and is transferred to the atmosphere in the form of water vapor, i.e., evaporation. This amount of water can be very significant and is not used in other economic activities. Thus, evaporation is taken as the amount of water consumed in the hydropower plants, and its calculation has been a challenge. The WFP of a given hydropower plant has been calculated using average meteorological datasets from the closest station, which has resulted in a biased calculation, mainly because of the changes in the climate over the last 20 years. WFP estimates are based on the ratio between evaporation from the reservoir and the electric energy produced by the facility on an annual/monthly time scale. In this case, it is known as gross water footprint (GWFP). However, WFP can also be calculated by considering net evaporation, referred to as the net water footprint (NWFP). In this chapter, the authors did present a case study involving a particular facility known as the Funil Hydropower Plant (FHP) in the Grande River basin, southeast Brazil, in which the authors of this chapter calculated WFP over two consecutive hydrological years using observed weather datasets recorded by four meteorological stations in the reservoir's surroundings, and the temperature of the reservoir's water, for including its thermodynamic behavior in the evaporation estimates. As per the findings, the annual GWFP of the Funil Hydropower Plant was 29 $m^3.GJ^{-1}$, while the NWFP was 13 $m^3.GJ^{-1}$, indicating a 55% reduction when adopting the latter methodology. This difference occurred due to the methodology used to calculate GWFP and NWFP. NWFP calculation considers the difference between evaporation and evapotranspiration from the flooded area before the reservoir filling. The reported results have demonstrated a reasonable efficiency of the water use for electric energy generation in this facility since both WFPs are low, allowing for characterizing a positive balance between water availability and water demand in the basin where the plant is located. In addition, the weather datasets are used in the reservoir's area and in the thermodynamics. Yet, the results showed efficiency of water use for electric energy generation in the Funil Hydropower Plant due to the small WFP, even when considering the GWFP as a reference, compared to other studies conducted in Brazil and worldwide. It has been found that the WFP of the Funil Hydropower Plant may be considered sustainable because of the positive water budget between availability and demand for water resources in the basin where the plant is installed.

The next chapter deals with the evaluation of the water footprint of the use and maintenance of buildings with a special focus on application to dwellings in Andalusia, Spain. Construction companies can both achieve financial success and reduce environmental impacts by carefully managing project budgets and considering the economic and environmental impacts simultaneously. This chapter evaluates the water footprint of the maintenance-and-use stage of buildings' life cycle, based on information in the project budget, using the computational tool PREDICE. In this study, the water footprint of the maintenance-and-use stage of buildings is evaluated based on the information in the budgets of the construction projects. To this end, the computational tool PREDICE is employed. The tool forecasts indirect environmental impacts by quantifying the embodied impacts in construction materials and machinery through life cycle analysis. This chapter also evaluates the direct impact derived from the water consumption of the occupants based on reported data. The chapter explores residential building construction in Andalusia, Spain, using the construction cost database and work breakdown system of the region. The construction project, specifically its bill of quantities (BoQ), is analyzed to predict maintenance tasks throughout its life cycle. Ninety-five maintenance work units and their cost are defined and integrated in the PREDICE tool. Two actual projects are studied and reported in this chapter: a private single-family dwelling and a 10-storey multi-family building that provides social housing. The results demonstrate how the use of the tool makes it possible to locate the sources with the greatest impact throughout the life cycle, thereby facilitating decision-making regarding the reduction of economic and environmental costs. The water footprint of the dwellings consists of 80% direct consumption by dwellers while 20% is due to the water embodied in construction materials and energy production. Also, water embodied in energy production is significantly as high as the construction material footprint. This tool quantifies embodied water in construction materials from life cycle analysis. The chapter also evaluates the direct impacts of occupants' water consumption, based on statistical data. It focuses on residential buildings in Andalusia and works with the region's construction cost database. The construction project budget is analyzed to predict maintenance tasks throughout its life cycle, defining the maintenance work and their cost in the PREDICE tool. Two actual projects are studied and reported in this chapter: a private single-family house and a 10-storey multi-family building for social housing. The results demonstrate how the tool facilitates decision-making to reduce both economic and environmental costs by locating the sources with the greatest impact throughout the life cycle. Environmental impact analysis of two types of building projects in Andalusia, a private single-family dwelling and a 10-storey multi-family block of social housing, was carried out throughout their life cycle. The construction phase was considered, as was the necessary maintenance after 20, 40, and 70 years. The construction stage and the maintenance when the building is 70 years old constitute the stages with the greatest economic impact and water footprint of the entire cycle.

When discussing water and sustainability, it is inevitable to address water treatment, which is covered in the final chapter, Sustainable Brine Management and Carbon Dioxide Capture: Minimizing Water Footprint, Addressing Environmental Challenges, and Promoting Resource Recovery. Desalination plants play a crucial role in meeting the growing demand for potable water. While they are highly efficient in producing desalinated water, they also generate a significant amount of reject brine, a highly concentrated saline waste stream. Disposing of this brine poses a major challenge for the desalination industry, as traditional methods often involve discharging it into the sea, resulting in environmental damage and marine pollution. Furthermore, this discharge amplifies the water footprint of desalination operations, exacerbating the overall environmental impact. The discharge brine can affect nearby water sources, like rivers, lakes, or underground aquifers. It makes the water in those sources less usable for things like drinking, agriculture, or supporting aquatic life. Additionally, the reliance on desalination in some countries contributes to greenhouse gas emissions due to the energy-intensive nature of the process.

This chapter aims to address these challenges and explore sustainable solutions for brine management in desalination. The increasing capacity of the desalination industry, driven by population growth and the depletion of other water resources, calls for urgent action. The primary concern lies in the emissions of pollutants, particularly brine, which has significant adverse effects on marine life. Furthermore, the continuous disposal of brine elevates seawater salinity, increasing energy requirements for desalination processes. Additionally, the energy-intensive nature of desalination plants, often located near power stations, results in large quantities of carbon dioxide (CO_2) emissions. In this chapter, the authors focus on highlighting brine management techniques, specifically electrodialysis using carbon dioxide. This approach enables the simultaneous removal of liquid and gaseous pollutants, while also offering the potential for creating value-added products such as carbonate/bicarbonate salts and inorganic acids. Additionally, the authors of the chapter did discuss the importance of pretreatment using selective electrodialysis to prevent scaling issues caused by the reaction of divalent ions with carbonate/bicarbonate ions.

By exploring these innovative brine management techniques, this chapter provides insights into sustainable practices that can help mitigate the environmental impact of desalination. Furthermore, it emphasizes the need for a holistic approach that addresses both liquid and gaseous pollutants, ensuring the long-term viability of desalination as a water supply solution. This chapter presents findings that support the concept of waste-to-resource generation and offers sustainable technology for combined CO_2 capture and reject brine management. This technology enables multi-waste handling with high-purity by-products, providing incentives for industries to consider waste as a renewable resource that can generate profits through the sale or use of by-products as raw materials. The presented technology has the potential to motivate industrialists to adopt sustainable waste management practices and contribute to the development of a circular economy.

"Sustainability and Water Footprint: Industry-Specific Assessments and Recommendations" offers a comprehensive exploration of water footprint concepts, evaluation methods, and strategies for reducing water use in various sectors. By examining case studies and innovative approaches, the book provides valuable insights for researchers, policymakers, and practitioners aiming to achieve sustainable water management. As readers delve into each chapter, they will gain a deeper understanding of the challenges and opportunities associated with water footprint assessment and learn how to contribute to a more sustainable future.

I take this opportunity to thank all the contributors for their earnest efforts to bring out this book successfully. I am sure the readers of this book will find it very useful.

With best wishes,
Dr. Subramanian Senthilkannan Muthu.

Industrial Grey Water Footprint: Principles, Evaluation Method, and Challenges

Keivan Arastou, Ali Dehnavi, and Shervin Jamshidi

1 Background and Definitions

1.1 Industrial Definition and Implementation

Industrial activities encompass the mass production of goods, often within specialized facilities known as industrial sectors. The term "industrial" relates to the systematic, large-scale manufacturing processes that characterize these production operations (Rumelt 1991). Implementation in this context refers to the practical application of industrial methods and technologies. It involves setting up systems, machinery, and workflows to achieve efficient production (Henderson et al. 1995). Ultimately, the goal is to create products that meet market demands while optimizing resources and minimizing waste. Industrial activities not only involve individual enterprises but also extend to collaborative groups categorized by sector or function. These groupings, often within industrial zones, facilitate shared resources and collective growth. Assessing both single entities and conglomerates allows for a comprehensive understanding of industrial dynamics, fostering development and innovation at every level.

1.2 Definition of Industrial Park

Industries are defined by their production and technological processes, significantly impacting economic development and environmental challenges within industrial

K. Arastou · A. Dehnavi · S. Jamshidi (✉)
Department of Civil Engineering, University of Isfahan, Isfahan, Iran
e-mail: keivan.arastou@gmail.com; a.dehnavi@eng.ui.ac.ir; sh.jamshidi@eng.ui.ac.ir

S. S. Muthu (ed.), *Sustainability and Water Footprint*, Environmental Footprints and Eco-design of Products and Processes, https://doi.org/10.1007/978-3-031-70810-7_2

ecosystems. The Global Industry Classification Standard (GICS) is an industrial classification that was established in 1999. The GICS structure consists of 11 sectors, 24 industry groups, 69 industries, and 158 sub-industries (MSCI S 2020). These industries can operate individually or collectively within an industrial park. Industries and their collective groups drive production, shaping the economy. Industrial parks support this by offering a hub for operations. Furthermore, these zones contribute significantly to pollution generation. In recent years, the global community has become increasingly alarmed by the depletion of resources and environmental degradation caused by rapid industrialization (Appannagari 2017). This awareness has led industries to pivot towards sustainable development, recognizing that the ongoing depletion of resources and discharge of various pollutants have led to significant threats to human well-being (Wang et al. 2019). Through industrial parks, manufacturers can share transportation and utilities, which fosters collaboration and optimizes the utilization of resources (Guteta and Worku 2023). The synergy between different sectors not only streamlines production but also fosters innovation. These areas are crucial for regional economic development, offering businesses a centralized area to operate efficiently (Sosnovskikh 2017).

Industrial parks are specialized zones designed to house numerous industries, providing them with the necessary infrastructure and facilities to operate efficiently. The advantages of industrial parks include the potential for economic growth through job creation and business attraction, streamlined access to utilities and services, and the possibility of improved environmental management due to centralized control (Table 1). However, there are also disadvantages to consider, as mentioned in Table 1 (Chen et al. 2019; Xueyi and Meyer 2011). Industrial parks can become hotspots for pollution, contributing to environmental impairment. The

Table 1 Advantages and disadvantages of industries and industrial parks (Chen et al. 2019; Xueyi and Meyer 2011)

Advantages	Disadvantages
Job creation and business attraction boost the local economy.	Pollution from centralized industries can harm the environment.
Provides essential services and facilities for efficient operation.	The influx of businesses can lead to increased traffic and noise.
Easier environmental and operational control.	Development can consume green spaces and agricultural land.
Can foster innovation through industry collaboration.	High demand for water, energy, and other resources.
Enhances the region's appeal to international investors.	Can alter the social fabric and displace local communities.
Centralizing industries can lead to more efficient use of resources and lower individual carbon footprints.	Industrial parks can concentrate pollutants, affecting air and water quality.
Opportunities for businesses to collaborate on environmental initiatives, like waste recycling and reuse.	Development can disrupt local wildlife habitats and ecosystems.

influx of businesses can also lead to increased traffic congestion, affecting the quality of life for nearby residents. Furthermore, the development of industrial parks often requires changing land use and land cover, which can result in the loss of green spaces and agricultural land, thereby altering the landscape and potentially deteriorating local ecosystems. Balancing these pros and cons is crucial for the sustainable development of industrial parks. These zones are specially designated areas for industries such as manufacturing plants and warehouses. These parks facilitate business operations by providing shared facilities and adhering to stringent zoning and environmental protection regulations. They strategically cluster different types of industries to minimize zoning conflicts. They are also designed for efficiency, housing a diverse array of industries from manufacturing to technology, and consequently, creating a dynamic environment (Felicio et al. 2016).

As shown in Table 2, industrial parks are increasingly focusing on sustainability and efficiency (Gourgiotis et al. 2021). They should continuously monitor and manage energy and water use, often utilizing renewable energy sources and setting targets to reduce greenhouse gas (GHG) emissions. Industrial waste management is also a chief concern although there are treatment systems in practice to recycle and reuse waste and wastewater. These parks strive to exceed national pollution control standards and use less hazardous materials. Location and social management are considered by preserving green spaces, assessing proximity to urban areas, evaluating economic impact, and ensuring accessibility to transport networks and social infrastructure. These measures reflect a commitment to environmental balance and sustainable development. Each category has indicators to measure the park's performance, aiming to promote environmental balance, resource efficiency, and social responsibility. The framework serves as a guide for industrial parks to operate sustainably and harmoniously within their ecological and social environments, as given in Table 2 (Gourgiotis et al. 2021).

Industrial parks provide easier access to essential services such as water, electricity, gas, transit roads, and human resources. Industrial park development is associated with two primary water concerns: (1) the stress of water withdrawal due to substantial freshwater consumption and (2) water pollution from the release of contaminants (Tian et al. 2014; Lyu et al. 2020). The effluents in these zones, originating from manufacturing processes, contain various chemicals or biological contaminants. Industrial wastewater varies greatly in both quality and quantity, depending on the type of industry and processes involved. It often contains a mix of inorganic and organic pollutants, suspended solids, chemicals, and sometimes toxic compounds like heavy metals (HMs), pesticides, and plastics (Munter 2003). The quantity and quality of wastewater generated by various industries can be substantial, containing a complex mixture of pollutants. This poses significant environmental challenges, including health risks, pollution prevention, and the need for effective treatment. Global cooperation is crucial for water quality preservation, and addressing these issues requires a holistic approach that combines prevention, treatment, and international efforts.

Table 2 A set of indicators to assess and monitor the transformation of industries and industrial parks (Gourgiotis et al. 2021)

Category	Description
Energy and water	Energy use at industrial parks is measured, focusing on the combined energy consumption.
	An industrial park uses renewable sources, which is an indicator of the park's use of renewable energy.
	An industrial park and its residents aim to reduce GHG emissions by setting carbon intensity targets.
Waste management and pollution control	Water use is monitored to manage demand, ensuring sustainable extraction levels.
	The number of equipped treatment facilities that can treat, recycle, and reuse wastewater, with the ratio of treated wastewater to total wastewater.
	An operational efficient system enhances water savings, with the ratio of water reused to total water consumed.
	An industrial park efficiency is reliant on material recycling and reuse by the ratio of reused solid waste to total waste.
	An industrial park targets less hazardous material use, with firm compliance as the measure.
	Ratio of waste material to landfill shows the ratio of properly treated to disposed waste in landfills.
	Encourage firms to not violate pollution control standards, with the standard adoption rate.
Industrial park location and social management	Open spaces in parks for environmental balance need to be preserved (more green space).
	Assess industrial park location by proximity to urban areas and financial capacity, using population data.
	Assess the regional economic impact using an index of exports and manufacturing value.
	Evaluation of industrial land plot sales and availability within receptors.
	Evaluation of industrial land availability based on depletion period.
	Consideration of proximity and accessibility to national transport networks.
	Consideration of cross-border features, infrastructures, and transport networks.
	Consideration of high- or medium-pressure gas pipelines in regional units.
	The regional units' access to highways, airports, ports, and railway stations.
	Distance categorization from educational and research institutions.
	An industrial park connected to social networks and public access.

1.3 Industrial and Industrial Park's Wastewater

Industrial wastewater is a complex mix of pollutants from factories that poses severe environmental problems with widespread impacts. Industrial wastewater comprises a diverse spectrum of pollutants, including chemicals, organic compounds, and HMs (e.g. Zn, Pb, etc.) carrying their ecological threats. This complex effluent is produced by different industries like textile and dyeing, food, metalworking, and other industries as shown in Table 3 (von Sperling and de Lemos Chernicharo 2005; Hanchang 2009; Barakat 2011; Muthu and Gardetti 2020; Huang et al. 2020; Ahmed et al. 2021). The metalworking industry tracks organic pollutants, toxic compounds (phenols), metals such as mercury, suspended solids (TSS), hydrocarbons, acidity (pH), and cyanide (CN). The food and dairy industry monitors organic waste, food particles, fats, oil, and grease (FOG) and maintains the balance of pH. Textile and dyeing operations keep tabs on dyes and chemicals (chemical oxygen demand (COD)), colorants, metals like Cr, wax, finishes, pH, and nutrients like nitrogen (N) and phosphorus (P). Pulp and paper processes measure paper fibers and other organics, processing chemicals, and pulp acidity. Non-ferrous metal industries monitor metal residues, chemical additives (e.g. phenols), metals like Cu and Zn, metal filings, lubricants, and acidic solutions. The chemical industry focuses on industrial chemicals, reactive agents (metals), chemical reactions, byproducts, and fertilizers. The electrical industry's concerns are solder, metals, and electroplating byproducts. Lastly, the plastic industry looks for plasticizers, stabilizers, catalysts, plastic granules, solvent mixtures, and colorants (von Sperling and de Lemos Chernicharo 2005; Hanchang 2009; Barakat 2011; Muthu and Gardetti 2020; Huang et al. 2020; Ahmed et al. 2021). Each industry's adherence to environmental regulations is crucial for minimizing the ecological damage.

For example, food and dairy industry produce wastewater with biochemical oxygen demand (BOD) and total suspended solids (TSS) ranging from 240 to 5900 mg/L and 60 to 5800 mg/L, respectively (Slavov 2017). Dairy production and milk-driven products result in the variation of BOD from 3000 to 5000 mg/L and total solids (TS) between 3000 and 7000 mg/L in industrial wastewater

Table 3 Wastewater pollutants based on the type of industries and industrial parks (von Sperling and de Lemos Chernicharo 2005; Hanchang 2009; Barakat 2011; Muthu and Gardetti 2020; Huang et al. 2020; Ahmed et al. 2021)

Industry	COD and BOD	Phenols	Metals	TSS	FOG	pH	CN	N and P
Metalworking	✓	✓	✓	✓	✓	✓	✓	–
Food and Dairy	✓	–	–	✓	✓	✓	–	–
Textile and Dyeing	✓	✓	✓	–	✓	✓	–	✓
Pulp and Paper	✓	–	✓	✓	–	✓	–	–
Non-ferrous metals	✓	✓	✓	✓	✓	✓	–	–
Chemical	–	✓	✓	–	–	✓	✓	✓
Electrical	–	–	✓	–	–	–	✓	–
Plastic	✓	✓	✓	✓	–	✓	–	✓

(Sirianuntapiboon et al. 2005). The chemical composition of whey and cheese exhibits significantly higher levels, with BOD at 20,000 mg/L and TSS at 22,000 mg/L (Deshpande et al. 2012). The presence of varying levels of impurities and pollutants in industrial wastewater is apparent. For instance, the characteristics of beverage industrial wastewater for BOD and TSS parameters differ between 728 and 1745 mg/L and 367 and 2940 mg/L, respectively (Amuda et al. 2006). Within a single industrial sector, the variations in water quality parameters can be substantial. For instance, in textile effluents, BOD concentrations may range between 80 and 6000 mg/L, while TSS can vary between 15 and 8000 mg/L (Yaseen and Scholz 2019). Characterization values for different textile wastewaters are diverse, like BOD and TSS for cotton fabric in the dyeing process, which ranges from 970 to 1460 mg/L and 120 to 190 mg/L, respectively, while in the bleaching process, BOD varies between 90 and 1700 mg/L and TSS ranges between 130 and 25,000 mg/L (Bisschops and Spanjers 2003). For instance, textile and chemical industries contribute to environmental degradation by discharging hazardous pollutants through wastewater, posing threats to both the ecosystem and public health (Khan and Malik 2014). Untreated industrial wastewater has adverse environmental impacts that jeopardize biodiversity and raise concerns about drinking water safety. Accordingly, the global approach is to urge industrial practices to take more environmental responsibility, by emphasizing the application of advanced wastewater treatment plants (WWTPs) through strict regulations. This approach highlights the requirements of some balance between industrial progress and environmental sustainability. It can encourage managers and supervisors to commit to more innovative solutions for preserving natural resources and biodiversity. For more comprehensive evaluation and monitoring, some recent studies have recommended using sustainability assessment indicators, like water footprint (WF).

The pressure of industrial parks on the surrounding environment increases due to the use of abstracted water and release of pollutants into the environment and water bodies. The concentration of pollutants in the raw industrial wastewater of industrial parks is highly variable and influenced by pollution from diverse industries. For instance, the concentration of COD (as an example), as illustrated in Table 4 (Chang et al. 2007; Fanchiang et al. 2009; Lei et al. 2010; Baban 2013; Zhu et al. 2015; Babatsouli et al. 2015; Liu and Tang 2018; Bay et al. 2018; Hu et al. 2020), at the outlet of industrial parks, depends on the category of their industrial groups.

The stress exerted by discharged wastewater may affect either the same water source from which it was withdrawn or a new destination where it is discharged into a receiving water body. Withdrawing water from a catchment for production in an Industrial Park and subsequently discharging industrial wastewater into a receiving water body or the same catchment can increase stress on the environment, amplifying associated concerns. For evaluating the performance of industries, a comprehensive sustainability index is essential, merging environmental and production metrics through a holistic view. This dual perspective ensures industries advance sustainably, respecting ecological boundaries while maintaining productivity. Assessing the destination of the discharge is crucial for understanding its adverse impacts on the environment. Moreover, the footprint of production per emitted pollution and

Table 4 Concentration of COD (mg/L) in raw wastewater of industrial parks

Row	Industrial park	Country	COD	References
1	Science-based industrial park	Taiwan	76.7–79.9	Chang et al. (2007)
2	Complex industrial park	Taiwan	278–627	Fanchiang et al. (2009)
3	Pharmaceutical industrial park	China	150–500	Lei et al. (2010)
4	Complex industrial park	Turkey	809–1095.5	Baban (2013)
5	Complex industrial park	China	154.7	Zhu et al. (2015)
6	Complex industrial park	Cyprus	730–7600	Babatsouli et al. (2015)
7	Chemical, pharmaceutical, and herbicidal industrial park	China	1200	Liu and Tang (2018)
8	Food and dairy industrial park	Iran	1090–2406	Bay et al. (2018)
9	Eco-industrial park	China	1000–6000	Hu et al. (2020)

water consumption should be considered for the final destination of discharges. Since balancing the ecological drawbacks with production gains is crucial for sustainability, the application of WF, and particularly grey WF (GWF), is pivotal, for highlighting environmental impacts alongside production benefits. This index can quantify pollution levels, reflecting the industry's water quality management under the production rate.

1.4 *Water Footprint (WF)*

The WF is an index that determines the embedded water consumed in an activity, directly and indirectly (Hoekstra et al. 2011). This index has three components: green, blue, and grey WFs (Fig. 1). The green WF (GrWF) represents the water consumed from precipitation or stored as soil moisture. On the other hand, a blue WF (BWF) is a variable that shows the surface and groundwater used directly in the production process or commodity. The grey WF (GWF), as the third variable of the WF index, represents the hidden (indirect) and measurable (direct) volume of freshwater needed to assimilate the discharged pollution loads based on the natural concentrations of the receiving water body (C_{nat}) to reach the allowable standard limit of pollutants into the environment (C_{max}). Determining the WF index needs three steps in order: (1) determining the scope of the study; (2) calculating the WF variables based on GrWF, BWF, and GWF; and (3) evaluating the cumulative results. Accordingly, to calculate WF, various factors are required, for formulating and adjusting the GWF for released pollution. However, before any calculation, the addressed study area should be defined at the beginning as a "study area". Determining this area specifies which variables should be included or excluded in the calculation of this index. Moreover, for GWF, C_{max} is determined based on the destination of the effluent or the required water quality level of the study area. One of the affected factors is the type of activity we are addressing. The WF can be

Fig. 1 Water footprint components; green (GrWF), blue (BWF), and grey (GWF). (By authors)

attributed to an agricultural crop, manufactured industrial product, or service. According to the estimations of Mekonnen and Hoekstra (2011), the total global WF of agricultural, industrial, and domestic activities was about 9087 billion cubic meters (BCM) in 2005 (Mekonnen and Hoekstra 2011a). Here, GWF constituted about 15% of the total WF, while the rest belonged to GrWF (74%) and BWF (11%). In addition, the overall share of agricultural, industrial, and domestic sectors in the total WF was calculated at about 81.5%, 15%, and 3.5%, respectively (Mekonnen and Hoekstra 2011b).

1.5 *Grey Water Footprint (GWF)*

The definition of an equivalent index for water pollution, based on the volume of freshwater needed for dilution, is not a new proposal. Previously, in 1974, Falkenmark and Lindh proposed that the amount of water needed to dilute pollutants in urban wastewater typically ranges from 10 to 50 times the volume of wastewater (Falkenmark and Lindh 1974). In 1996, Postel and colleagues stated a general dilution factor (28 L/s per 1000 people) for pollution abatement in municipal wastewater to a certain concentration of contamination (Postel et al. 1996). According to the general dilution factors in previous methods, the type and amount of pollution along with the required treatment level were not specified. For this reason, Chapagain et al. (2006) presented the dilution factor considering the type and amount of pollutants, based on water quality standards in a case study (Chapagain et al. 2006). For the first time in 2008, Hoekstra and Chapagain introduced the term "grey water footprint" as the volume of water needed to reduce the pollution load to the permissible threshold of each pollutant (Chapagain and Hoekstra 2008). Afterward, GWF has been defined by the quantity of water necessary to dilute pollutants to a level that ensures that the water quality maintains compliance with ambient water quality standards (Franke et al. 2013).

The GWF_i index for a pollutant (i) can be determined using the contaminant concentration (C_i) and effluent flow rate (Q) as follows, in Eq. 1 (Hoekstra et al. 2011):

$$\mathrm{GWF_i} = \left(\frac{\mathrm{LOAD}}{C_{\max} - C_{\mathrm{nat}}}\right)_{\mathrm{i}} = \left(\frac{Q \times C}{C_{\max} - C_{\mathrm{nat}}}\right)_{\mathrm{i}} \tag{1}$$

The presence of the pollution load (LOAD) is the amount of pollution added to the receiving environment, like a watershed, and it is not the pollution load already present in the receiving water body. For calculating the GWF, the pollution load is divided by the difference of the maximum acceptable concentration (C_{max}) and the natural concentration in the water body (C_{nat}). This gap ($C_{max} - C_{nat}$) is called the maximum permissible addition (MPA) of the ambient water and represents the permissible level of adding pollution, based on the maximum allowable concentration of receiving water bodies and the natural concentration of contaminants. Yet, this index depends on a set of critical variables for its final value. The GWF is typically calculated using Eq. 1. However, accurately estimating the GWF involves several details, which are introduced in the subsequent section.

1.5.1 Agricultural and Domestic GWF

The application of GWF is categorized into three groups: agriculture, industry, and domestic. This index has been used in literature for reporting agricultural activities as well as domestic, as summarized in Table 5 (Hoekstra and Mekonnen 2012; Mekonnen and Hoekstra 2015, 2018; Morera et al. 2016; Gu et al. 2016; Hu et al. 2018; Borsato et al. 2018; Karandish 2019; Zhang et al. 2019; Jamshidi et al. 2020, 2022, 2024; Stejskalová et al. 2021; Roudbari et al. 2023). The table presents research findings on GWF, emphasizing the major role of agriculture on the GWF of products, the reduction potential through improved agricultural strategies, and the efficacy of wastewater treatment in reducing the GWF. It also points out the rising GWF trend due to urban wastewater and specific GWF measurements in rice production and other crops. Monitoring specific pollutants, like total phosphor (TP) and total nitrogen (TN), is essential for managing the GWF of particular crops. Generally, these studies highlight the importance of sustainable practices in agriculture and wastewater treatment to address the challenges of water scarcity and environmental conservation.

Table 5 The summary of the outcomes of studies conducted on the agricultural and domestic GWF

Row	Main finding	References
1	Hoekstra and Mekonnen (2012) reported that the global WF for agricultural product production was 8362 BCM/year, with a GWF proportion of about 9%.	Hoekstra and Mekonnen (2012)
2	Mekonnen and Hoekstra (2015) found that the global GWF from anthropogenic nitrogen amounted to 13,000 BCM/year, in which agriculture and industry contributed 75% and 2%, respectively.	Mekonnen and Hoekstra (2015)
3	Morera et al. (2016) showed that a municipal wastewater treatment plant with an activated sludge process could reduce the GWF due to nitrogen and phosphorus by 85.3% and 70.8%, respectively.	Morera et al. (2016)
4	Gu et al. (2016) showed that the A_2O treatment process had the highest efficiency in reducing the GWF from municipal wastewater, compared to the least efficient trickling filter process.	Gu et al. (2016)
5	Hu et al. (2018) reported that China's GWF from agricultural drainage was 2598 BCM/year, and 77% of the overall GWF was due to chemical fertilizers.	Hu et al. (2018)
6	Borsato et al. (2018) determined that the GWF from agricultural production could be reduced by 10%–11% through the adjustment of chemical fertilizers and pesticides.	Borsato et al. (2018)
7	Mekonnen and Hoekstra (2018) estimated the global GWF from anthropogenic phosphorus to be 147,000 BCM/year, with 38% from agriculture and 8% from industry.	Mekonnen and Hoekstra (2018)
8	Karandish (2019) noted that approximately 71.9% of Iran's GWF index from agricultural drainage was related to grain production.	Karandish (2019)
9	Zhang et al. (2019) reported in 2019 that the overall GWF in China increased by 16.7% between 2003 and 2015, when 70% of the increased GWF was attributed to municipal wastewater.	Zhang et al. (2019)
10	Jamshidi et al. (2020) compared the impacts of farm management practices on the GWFs of crops in a lake basin through a soil and water assessment tool (SWAT). Here, TN and TP were the most critical pollutants and GWF contributed to about 4% of total WF.	Jamshidi et al. (2020)
11	Stejskalová et al. (2021) assessed the GWF due to parameters like NH_4 and TP, which were calculated to be 3.2 and 0.547 BCM/month, respectively.	Stejskalová et al. (2021)
12	Jamshidi et al. (2022) developed and used GWF for the sustainability assessment of agricultural productions on a basin scale.	Jamshidi et al. (2022)
13	Vaez Roudbari et al. (2023) determined the GWF of paddy rice, influenced by various parameters, such as TP and TN, to be 6055 m^3/ton, as mainly measured by the TP parameter.	Roudbari et al. (2023)
14	Jamshidi et al. (2024) used GWF as a multi-pollutant index for evaluating the performance of domestic WWTPs more comprehensively within the life cycle assessment (LCA) framework. Here, GWF accounted for about 87% of the total WF of WWTPs and reduced to 67% by reuse.	Jamshidi et al. (2024)

It should be added that the GWF in agricultural products has generally been less explored in comparison with the other two WF variables (BWF and GrWF). Therefore, there are still some gaps for more research and case studies on GWF modification, which can contribute to improving agricultural indicators.

1.5.2 Industrial and Industrial Park's GWF

GWF emerges as a practical factor for assessing the regional environmental impacts of industrial production on ecosystems. It specifically addresses the pollution levels in specific freshwater resources (Weerasooriya et al. 2021). GWF in industrial settings originates from various processes within manufacturing facilities, including cooling, cleaning, polishing, fermentation, and production-related tasks like dyeing (Mousazadeh et al. 2021). Managing this wastewater is critical due to potential environmental risks through discharging multiple pollutants (such as HMs, chemicals, and organic compounds) into the environment (Munter 2003). Industrial companies often bear responsibility for contributing to elevated water pressure, excessive water consumption, and larger GWF. These challenges are not unexpected, especially considering that the manufacturing industry is projected to significantly increase its water demand by 2050 by four-fold compared to 2000 (Leflaive et al. 2012). GWF is a comprehensive metric index that evaluates how industrial activities affect water quality, as shown in Table 6 (Zhang et al. 2019; Wang et al. 2013; Handayani et al. 2019; Herrebrugh 2018; Yapıcıoğlu 2019a, b; Ansorge et al. 2020a, b, c; Dong et al. 2021; Yan et al. 2021; Meng et al. 2022).

The GWF has recently become a critical indicator for comprehensive industrial impact assessment, particularly on water resources. However, studies across the globe have revealed that these GWFs have significant variations. Herrebrugh (2018) estimated 31,900 BCM/year of industrial GWF globally, emphasizing the pervasive nature of this issue (Herrebrugh 2018) and necessitating worldwide collaborative efforts for using sustainable solutions. The estimated GWF for industrial activities depends upon different sectors like chemicals, metals, textiles, etc., all over the world. Wang et al.'s (2013) analysis of China's textile industry unveiled a high GWF of 62.34 BCM/year due to the COD parameter (Wang et al. 2013), signaling an urgent need for pollution mitigation strategies. On a smaller scale, Handayani et al.'s (2019) research contrasted this result with Indonesia's textile sector, where natural and synthetic dyes resulted in two GWFs based on COD about 23,800 m^3/year and 101,178 m^3/year, respectively (Handayani et al. 2019). The latter highlights the environmental benefits of natural materials for pollution control. Zhang et al. (2019) further dissected China's textile industry, attributing 14.4% of the nation's industrial GWF to this sector (Zhang et al. 2019), underscoring its significant environmental footprint. However, GWF is not solely attributed to one industry. It can also stem from various sectors, for example, electricity generation facilities and petrochemical industries. These challenges significantly contribute to the overall environmental impacts through emissions and resource consumption.

Table 6 The summary of the conditions and results of studies conducted on the industrial GWF

Row	Main finding	References
1	Wang et al. (2013) reported that the GWF of the textile industry's wastewater in China was about 62 BCM/year.	Wang et al. (2013)
2	Handayani et al. (2019) found that in Indonesia, the GWFs of textile industrial wastewater, due to natural and synthetic dyes, were 23,800 and 101,178 m^3/year, respectively.	Handayani et al. (2019)
3	Herrebrugh (2018) estimated the global GWF from industrial wastewater at 31,900 BCM/year.	Herrebrugh (2018)
4	Zhang et al. (2019) reported that the textile industrial proportion of China's GWF from all industrial wastewater was about 14%.	Zhang et al. (2019)
5	Yapıcıoğlu's (2019a) study estimated the GWF of dairy industrial wastewater in Turkey, measured using the TSS parameter, to be approximately 41,026 m^3/day.	Yapıcıoğlu (2019a)
6	Yapıcıoğlu (2019b) found that the GWF from the painting industrial wastewater in Turkey, measured by the COD parameter, was about 1129 m^3/month.	Yapıcıoğlu (2019b)
7	Ansorge et al. (2020a) calculated the average GWF from industrial wastewater in the Czech Republic, using the NH_4 parameter, to be about 5.13 BCM/year.	Ansorge et al. (2020a)
8	Ansorge et al. (2020b) reported that the GWF from the Temelín Nuclear Power Plant's wastewater in the Czech Republic, measured by tritium pollution, was about 95 MCM/month.	Ansorge et al. 2020b)
9	Ansorge et al. (2020c) calculated the average total GWF in the Czech Republic, using the NH_4 parameter, to be 21.78 BCM/year.	Ansorge et al. (2020c)
10	Dong et al. (2021) determined the GWF from the wastewater of an industrial park in China, using the TP parameter, which was about 1.07 MCM/year.	Dong et al. (2021)
11	Yan et al. (2021) reported that the GWF from non-carcinogenic HMs in the wastewater of an industrial park in China, specifically from copper pollution, was 1.4 BCM/year.	Yan et al. (2021)
12	Meng et al. (2022) reported that the GWF from industrial wastewater in the city of Yantai, China, was reduced to 46 MCM in 2019, compared to 116 MCM in 2015.	Meng et al. (2022)

Ansorge et al.'s (2020) findings in the Czech Republic showed that the average industrial GWF of this country was about 5.13 BCM/year (maximum GWF due to NH_4 was determined to be 6.53 BCM/year). The GWF of the Temelín Nuclear Power Plant in the Czech Republic was 94.7 MCM/month due to tritium pollution (Ansorge et al. 2020a; Ansorge et al. 2020b), reflecting the diverse industrial influences on water quality. Food and dairy products constitute a significant portion of the global industrial GWF. Yapıcıoğlu (2019) estimated industrial GWF based on dairy and paint industries, finding values of approximately 41,026 m^3/day for TSS and 1129 m^3/month for COD, respectively (Yapıcıoğlu 2019a, b), thereby demonstrating the ongoing challenges of multiple pollutants in industrial wastewater management.

Dong et al. (2021) reported that a Chinese industrial park had a GWF for TP of approximately 1074 MCM/year (Dong et al. 2021), suggesting some success in pollution reduction and control. For this area, various pollutants, such as HMs, must be monitored and examined continuously for GWF assessment. Yan et al. (2021) focused on non-carcinogenic HMs, particularly Cu, with GWF of about 1.4 BCM/year (Yan et al. 2021), pointing to the critical need for addressing toxic pollutants. These assessments, on a regional or national scale, demonstrate the long-term performance of policies and planning in the management and reduction of pollution. Meng et al. (2022) revealed GWF reduction in Yantai City, China, from 116 MCM/year in 2015 to 46 MCM/year in 2019 (Meng et al. 2022), offering a beacon of progress, showcasing the effectiveness of enhanced pollution control approaches. In general, these studies underscore the imperative of vigilant GWF assessment and management to safeguard water resources and promote environmental sustainability, as each region and industrial area have unique challenges and opportunities for improvement.

The GWF is an essential metric in the contemporary monitoring of industrial and agricultural wastewater. It evaluates a multi-parameter aspect, considering various pollutants and emphasizes the pollution load to gauge environmental impacts. GWF's comprehensive approach allows for integration into management strategies, aiding wastewater treatment and reuse. However, GWF has its challenges. Data scarcity can hinder comprehensive assessments, and the complexities in interpreting GWF values for different pollutants may require specialized expertise. These issues, along with the need for standardization in measurement methods, present gaps that must be addressed. The discussion continues to explore gaps and challenges, aiming to refine GWF's application in sustainable water management. In addition, the assessment of GWF for composite industrial effluents, which are characteristic of diverse enterprises operating within industrial parks, has not been extensively studied. This oversight is significant because such wastewater often contains a myriad of contaminants. Therefore, a more nuanced and detailed analysis is necessary to accurately quantify the GWF in these contexts, taking into account the variable nature of the effluents and the specific environmental standards applicable to the region.

This chapter introduces methods that can be adapted not just for large-scale industrial operations but for single factories as well. Implementing these flexible strategies can lead to improved efficiency and sustainability in any industrial setting. Based on the definitions mentioned earlier, in this chapter, the authors first discuss about the gaps and challenges of industrial GWF. Second, this chapter introduces the evaluation methods, calculation steps, and the required scenarios for GWF accounting. Here, we argue about specific details that researchers should consider in future assessments. Finally, the industrial GWF of a study area in Isfahan, Iran, was quantified based on genuine sampled data. Here, an assessment framework for GWF is strategically designed to encompass a variety of scenarios, thereby facilitating a more precise GWF quantification. Furthermore, an analysis of the WF within an industrial region is conducted illustratively to gauge the industrial activities' impacts on the regional water body.

1.6 Gaps and Challenges

The assessment of the GWF presents several challenges and gaps that need to be addressed for a more comprehensive understanding of water quality management:

- The pollution emission should be first examined, and then, calculate the GWF based on examined data, rather than estimating or assessing emissions based on data from other locations. Given the direct impacts of the GWF variables, such as volumetric flow rate and pollutant concentration, as well as the high variations of these factors in industrial wastewater, it is strongly recommended to have accurate measurements. This fact is sometimes neglected for GWF assessment, particularly on large scales.
- The GWF is assessed by considering pollution loads discharged into water bodies. Unlike GrWF and BWF, which focus on water consumption and availability, the GWF cannot be directly calculated using a consumption or water-balance approach. Instead, it relies on an assessment that requires both the wastewater volume and pollutants' concentrations and simultaneously considers the unique characteristics of the receiving water body (Johnson and Mehrvar 2021). Due to these variations, comparing GWF between different facilities and productions is challenging. Therefore, continuously testing and analyzing the volumetric flow in both the inlet and outlet and multiple pollutants' concentrations can dramatically reduce GWF uncertainties and improve its accuracy.
- The net GWF should be calculated by considering the impacts of pollutant concentrations in consumed water from inlet (C_{act}) to outlet (C_{eff}), contrary to many recent studies (Yapıcıoğlu and Yeşilnacar 2022). This approach ensures that the measurement reflects the true net environmental impact of consumed water, rather than just the volume withdrew or discharged. It is also about assessing the quality of this water, not just the quantity for WF assessment.
- The determination of the GWF is often based on a limited number of parameters in studies; however, this amount can vary depending upon the inputs and outputs of each industry or treatment unit and may be influenced by different parameters. Therefore, the determination of this index should be based on multiple pollutants to estimate more accurate GWF based on the leading type of pollution.
- If the industrial effluents are reused for different purposes, like irrigation, any potential runoff or drainage should be considered in the assessment of industrial GWF. If the effluents are discharged at multiple points and locations, the overall GWF should be the accumulated GWF calculated for each discharge location. It is also noteworthy that recycling and reuse of treated industrial wastewater for self-production or environmental rehabilitation would have dramatic impact on the overall WF of the products. Yet, GWF and BWF assessments take an industrial activity as a black box and ignore any inner flows and material cycles, in contradiction with life cycle assessment (LCA).
- When assessing the net GWF of pollutants, a negative value indicates that the wastewater has a diluting effect, rendering the GWF index null for final determination. This scenario is exemplified by the water softening process in industrial

units, which can result in a negative GWF. Conversely, pollutants that remain unchanged or are produced during the treatment process may cause an increase in GWF. Nonetheless, industrial treatment facilities typically aim to maintain effluent concentrations (C_{eff}) below the environmental water quality targets (C_{max}), ensuring compliance with regulatory standards.

- For GWF accounting, one decisive factor is C_{max}. This variable can be considered based on various scenarios based on guidelines and standards for water reuse or wastewater discharge. It is worth mentioning that the selection of different scenarios for C_{max} is due to the potential impacts of various natural and anthropogenic events on the exposed area and the effect of pollutants on other locations. Choosing very strict and highly limited C_{max} values will result in higher GWFs. On the contrary, lenient and poorly limited C_{max} may lead to smaller GWFs. Therefore, determining the GWF indicator according to different scenarios of C_{max} helps in the accurate assessment of this indicator.
- Determination of C_{nat} constitutes one step in GWF accounting. Due to data scarcity in many locations or limited access to sufficient data, it is recommended that C_{nat} be determined based on globally constant values for certain parameters (Franke et al. 2013). It is important to note that if a global constant value is used for the natural concentration of pollutants, adjustments may be necessary based on regional assessments.
- One of the challenges of the GWF is calculating its removal rate, defined as GWF reduction efficiency (GWFRE), like when we utilize wastewater treatment units. In Gu's study (2016), this variable is calculated using Eq. 2 (Gu et al. 2016), while the correct GWF reduction efficiency (GWF%) in this study is introduced as Eq. 3 (by authors):

$$\mathrm{GWFRE} = \min\left(\frac{C_{in} - C_{out}}{C_{out}}\right)_i \times 100 \tag{2}$$

$$\mathrm{GWF\%} = \left(\frac{\mathrm{GWF}_{in} - \mathrm{GWF}_{out}}{\mathrm{GWF}_{in}}\right)_i \times 100 \tag{3}$$

- One of the significant challenges in determining GWF is accounting for multiple water extraction points for production and spatially variable points for wastewater discharge. Provided that water is abstracted from different resources (surface water and groundwater), the C_{act} should be separately compatible with the sources. For wastewater discharge, if there are multiple locations or receiving water bodies, the GWF must be determined separately for each environment and then assessed collectively to determine the overall GWF.
- The concentration of pollutants in the output varies according to the treatment method applied. Therefore, assessing the impact of these concentrations on the GWF is essential. This evaluation presents a challenge due to the diverse treatment processes. Moreover, this necessitates a comprehensive analysis to ensure sustainable water management practices that mitigate adverse ecological consequences.

- When determining the GWF based on multiple pollutants, a critical parameter is identified, and the GWF is determined based on the leading pollutant. However, it is important to note that pollution parameters can be weighted for GWF accounting based on their impacts or equivalent damages. Therefore, the previously critical parameter may no longer be critical with this perspective. It is also worth mentioning that weighting should not be confined to the concentration of pollutants, as it can also be applied to the variable of the GWF depending on the type and location of discharge or water abstraction.
- The concept of GWF is subject to uncertainties due to its reliance on quality-based measures, various influential parameters, and the assimilation capacity of the receiving water bodies. These uncertainties can pose challenges to the application of GWF and the establishment of effective policies. Furthermore, global standards may not perfectly represent the unique environmental conditions or the natural self-purification potential of local water bodies, whereas they offer a general framework that can be adopted to improve uniformity in GWF accounting (Jamshidi 2021).
- There is an additional aspect observed in some developing countries. An industrial wastewater that violates the minimum standards required for WWTPs, for example having a large amount of toxic materials, cannot be directly mixed with other raw industrial effluents in the inlet of this centralized WWTP. Instead, it is transported to another location for disposal via hazardous sewage tanker trucks. The GWF of this portion should be accounted for in a separate catchment area (discharged place). Assessment of the GWF at each location is based on the actual discharge.

2 Evaluation Method

2.1 Variables and Standards

The GWF is an important environmental metric that measures the amount of freshwater needed to dilute pollutants to acceptable levels. The variables such as C_{max}, C_{nat}, and C_{act} play a crucial role in its calculation. C_{max} represents the maximum concentration of a pollutant that is allowed in a water body, ensuring the safety of the ecosystem and human health. C_{nat} indicates the natural concentration of the pollutant in the water body without human interference, which is essential for understanding the baseline level of pollution. C_{act} is the actual concentration of the pollutant found in the intake water. The mentioned variables, combined with the flow rate of the abstracted water (Q_{abstr}), outlet wastewater (Q_{eff}), and the concentration of pollutants (C_{eff}) are essential for GWF accounting. By comparing these variables, we can assess the impact of human activities on water quality and calculate the GWF, which helps in making informed decisions for sustainable water management. This assessment is particularly sensitive to the specific regional conditions and the characteristics of the aquatic environment where the discharge occurs.

2.1.1 C_{max}

GWF index is determined based on the C_{max} of the receiving environment (Hoekstra et al. 2011). To establish C_{max}, various qualitative standards, including surface water quality regulations, drinking water quality standards, water and wastewater guidelines for irrigation and agriculture, recreational water quality criteria, etc., are utilized (Hoekstra et al. 2011; Franke et al. 2013). Selecting an appropriate standard depends on the destination of the effluents and the type of pollutants. Four scenarios are introduced in the study: extremely strict (Scenario 1), moderately strict (Scenario 2), moderately mild (Scenario 3), and significantly mild (Scenario 4) limitations, as detailed in Table 7 (Stejskalová et al. 2021; Dong et al. 2021; Department of Environment Islamic Republic of Iran 2016; OECD 2006, 2007; Department of Environment Islamic Republic of Iran 2023; Vice President's Planning and Strategic Supervision 2010; FAO 2023; Taiwan EPA 2022; National Health and Medical Research Council 2008; Ministry Environment And Water of Malaysia 2021; Health Ministry of New zealand Government 2019; Minister of Health Canada 2022; Swistock 2016; Fipps 2003). For instance, for selecting a threshold for wastewater discharge into the groundwater, the criteria would be based on the potable water quality standards or irrigation, while for surface water, it can be determined according to the ecological or recreational guidelines, in addition to potable water and irrigation (Hoekstra et al. 2011).

These standards set permissible limits for various pollutants in water and wastewater. C_{max} serves as a critical parameter in assessing the environmental impact of grey water discharge, as it establishes the threshold level beyond which the assimilation capacity of the receiving water body is compromised. The volume of water consumption required to assimilate the pollution is intricately linked to the threshold level determined by C_{max}. Proper adherence to C_{max} values is essential for sustainable water management, ensuring that the discharged grey water conforms to the established quality standards and minimizes the ecological strain on the receiving water bodies.

Due to the diverse impacts of effluent discharge on the receiving environment, the utilization of different water quality standards to determine C_{max} adds a comprehensive and scientifically rigorous dimension to the evaluation process. The GWF index becomes more sensitive when water quality control standards become stricter, and evaluating through a range of scenarios for C_{max} allows for a nuanced appraisal of this index (Cui et al. 2020). This approach enables the quantification of the GWF index where there are diverse endpoints for effluent discharge. The examples are given in Table 7 for C_{max}. For instance, the maximum allowable value for BOD varied between 5 (Department of Environment Islamic Republic of Iran 2016) and 100 mg/L (Department of Environment Islamic Republic of Iran 2023). This limit for Pb was determined between 0.01 (Department of Environment Islamic Republic of Iran 2016; Health Ministry of New zealand Government 2019) and 10 mg/L, as the latter was recommended for the constituents in reclaimed water for short-term irrigation in the United States (Fipps 2003). By employing this methodology, one

Table 7 An example of various maximum allowed concentrations of pollutants (Scenarios)

Leading pollutants	Unit	Scenario 1		Scenario 2		Scenario 3		Scenario 4	
		C_{max}	References	C_{max}	References	C_{max}	References	C_{max}	References
BOD	mg/L	5	Department of Environment Islamic Republic of Iran (2016)	7	OECD (2006)	30	Department of Environment Islamic Republic of Iran (2023)	100	Vice President's Planning and Strategic Supervision (2010)
COD	mg/L	7	OECD (2007)	30	OECD (2006)	60	Vice President's Planning and Strategic Supervision (2010)	200	Vice President's Planning and Strategic Supervision (2010)
P	mg/L	0.15	Stejskalová et al. (2021)	0.2	OECD (2007)	0.4	Dong et al. (2021)	1.2	OECD (2006)
PO_4	mg/L	0.9	OECD (2006)	2	FAO (2023)	6	Vice President's Planning and Strategic Supervision (2010)	50	Vice President's Planning and Strategic Supervision (2010)
NH_3	mg/L	0.3	Taiwan EPA (2022)	0.5	National Health and Medical Research Council (2008)	2	Dong et al. (2021)	2.7	Ministry Environment And Water of Malaysia (2021)
NH_4	mg/L	0.23	Stejskalová et al. (2021)	0.5	Health Ministry of New zealand Government (2019)	2	OECD (2006)	2.5	Vice President's Planning and Strategic Supervision (2010)
NO_2	mg/L	0.2	Health Ministry of New zealand Government (2019)	1	Minister of Health Canada (2022)	3	Health Ministry of New zealand Government (2019)	10	Vice President's Planning and Strategic Supervision (2010)
NO_3	mg/L	5	Swistock (2016)	10	Swistock (2016)	11.3	OECD (2006)	50	Vice President's Planning and Strategic Supervision (2010)

Cr	µg/L	50	Department of Environment Islamic Republic of Iran (2016)	100	Fipps (2003)	500	Vice President's Planning and Strategic Supervision (2010)	1000	Department of Environment Islamic Republic of Iran (2023)
Cu	µg/L	50	Vice President's Planning and Strategic Supervision (2010)	200	Department of Environment Islamic Republic of Iran (2023)	1000	Vice President's Planning and Strategic Supervision (2010)	2000	Health Ministry of New zealand Government (2019)
Pb	µg/L	10	Department of Environment Islamic Republic of Iran (2016)	1000	Vice President's Planning and Strategic Supervision (2010)	5000	Fipps (2003)	10,000	Fipps (2003)
Zn	µg/L	1500	Health Ministry of New zealand Government (2019)	2000	Vice President's Planning and Strategic Supervision (2010)	5000	Minister of Health Canada (2022)	10,000	Fipps (2003)

can ascertain the volumetric measure of freshwater needed to assimilate a specific load of pollutants, thereby reflecting the potential environmental impact of various water quality scenarios.

2.1.2 C_{nat}

The natural concentration of pollutants in the receiving water body (C_{nat}) encompasses the inherent levels of environmental substances observed over a specific temporal duration, elucidating dynamic variations influenced by natural processes. C_{nat} for any pollutant index represents a concentration solely attributable to natural processes and the geological characteristics of the region existing within the water body (Gaillardet et al. 2003). Consequently, C_{nat} serves as the concentration of pollutants in the receiving water environment, reflecting the concentration of each pollutant without any human intervention, essentially indicating the pollution level before industrialization in a specific region (Hoekstra et al. 2011; Franke et al. 2013). Contrary to the background concentration, typically denoting a steady baseline level at a specific spatial point in the water body, C_{nat} transcends the confines of a certain location, emphasizing a temporal dimension (Fig. 2).

The crucial distinction lies in its temporal scope, as C_{nat} scrutinizes variations over time, acknowledging its inherent variations in environmental conditions due to non-anthropogenic factors. It is essential to note that selecting the same C_{nat} for each pollutant globally is not recommended due to temporal and spatial variations in water quality. Therefore, referring to the local data to determine C_{nat} is recommended for accuracy and relevance (Directorate-General for Health and Food Safety (European Commission) 2018). Franke et al. (2013) highlighted that in the absence of suitable local data, using reliable global information may be challenging. However, it is necessary and useful for determining this variable (Franke et al. 2013). More importantly, for materials and compounds with anthropogenic origins, C_{nat} is considered to be zero. It should be remembered that if C_{nat} is incorrectly taken as zero, GWF will be incorrectly underestimated.

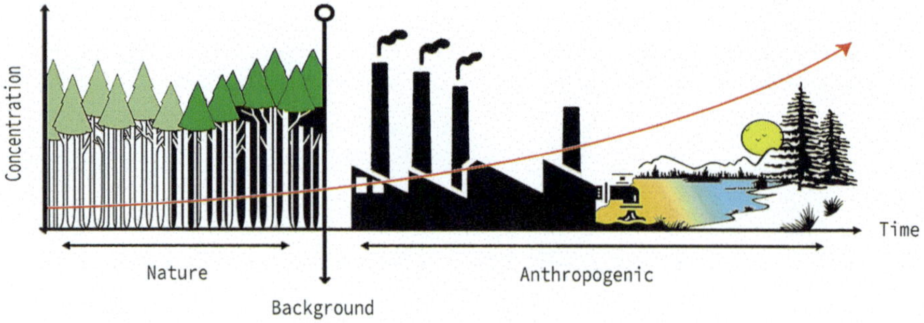

Fig. 2 Comparison of natural and background concentration (by authors)

2.1.3 C_{act}

The actual concentration of pollutants in abstracted water (C_{act}) serves as a useful variable in net GWF accounting. It accounts for the presence of impurities or pollutants in water before any application. The influence on GWF is more pronounced when contamination concentrations are higher in wastewater, as opposed to lower levels found in natural water sources. Notably, the practice of reusing treated wastewater for industries can lead to a substantial reduction in GWF.

2.1.4 D_f

The dilution factor is a pivotal dimensionless fraction in the environment, signifying the number of times an effluent must be diluted with the ambient water to reach a safe concentration level, as given in Eq. 4 (Hoekstra et al. 2011). Among multiple pollutants, the maximum calculated number is considered as D_f, which defines the dilution factor for all the other parameters that have a lower D_f indicator. This factor, in conjunction with the water body's assimilation capacity, refers to the natural ability to absorb and neutralize pollutants. D_f ensures that the effluent's concentration remains within the ecosystem's carrying capacity. By respecting both the dilution factor and the assimilation capacity, we can maintain the ecological balance and uphold sustainable water management.

$$D_f = \left(\frac{C_{eff} - C_{act}}{C_{max} - C_{nat}} \right)_i \tag{4}$$

2.2 GWF Calculation

2.2.1 General Methods

The GWF index includes the following variables (Eq. 5), magnitude of the discharged wastewater flow rate (Q_{eff}: volume/time), the flow rate of water abstraction (Q_{abstr}: volume/time), the concentration of discharged wastewater pollutants (C_{eff}: mass/volume), the concentration of impurities and contaminations of the abstracted water (C_{act}: mass/volume), the natural concentration of pollutants, related to the period before anthropogenic transformation (C_{nat}). In addition, the maximum allowable concentration of pollutants in the receiving water body (C_{max}) is a remarkably sensitive and another influential variable. The latter represents a safe threshold based on the exposure and risks of hazardous materials, the carrying capacity of the receiving environment, and the purpose of reusing treated wastewater, all three of which are considered when determining water and wastewater quality standards (C_{max}). The accounting for discharged pollution loads in the global WF

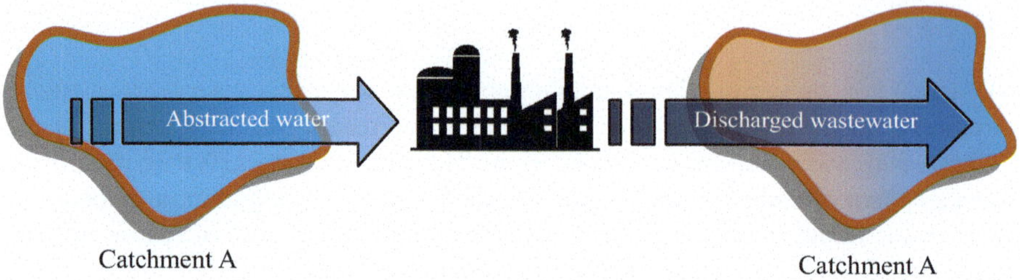

Fig. 3 Catchment A subjected to the water withdrawal and assimilation pressures (GWF). (By authors)

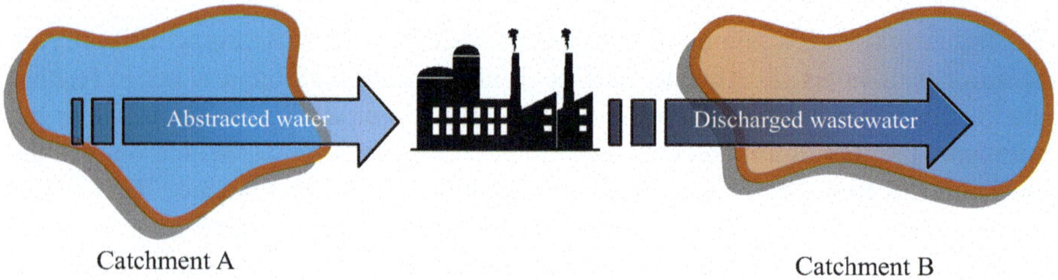

Fig. 4 Catchment B subjected to the assimilation pressures (GWF). (By authors)

framework is fundamentally based on three key elements: C_{max}, C_{nat} (the natural concentration), and C_{act} (the actual concentration). These variables will be elucidated in future sections. According to Eq. 1, the GWF of each pollutant, denoted as GWF_i, is used to evaluate the WF in a catchment subjected to two distinct pressures: water withdrawal and pollution dilution up to the permitted threshold (catchment A in Fig. 3).

In the context of catchment A shown in Fig. 5, the GWF_i is calculated using Eq. 5 (Hoekstra et al. 2011):

$$\mathrm{GWF_i} = \left(\frac{(Q_{\mathrm{eff}} \times C_{\mathrm{eff}}) - (Q_{\mathrm{abstr}} \times C_{\mathrm{act}})}{C_{\max} - C_{\mathrm{nat}}}\right)_{\mathrm{i}} \tag{5}$$

In the case of catchment B, as shown in Fig. 4, which is undergoing contaminant dilution while water withdrawal transpires in catchment A, it is noteworthy that the water flow rate of abstraction (Q_{abstr}) may be equal to the effluent flow rate (Q_{eff}).

In such instances, the calculation of GWF_i in catchment B follows Eq. 6 (Hoekstra et al. 2011). This equation offers a quantitative metric for GWF, intricately addressing the relationship between pollutant dilution in one catchment and water withdrawal in another.

$$\mathrm{GWF_i} = \left(Q_{\mathrm{eff}} \times \frac{C_{\mathrm{eff}} - C_{\mathrm{act}}}{C_{\mathrm{max}} - C_{\mathrm{nat}}} \right)_{\mathrm{i}} \tag{6}$$

Subsequently, to evaluate the GWF attributed to distinct pollution indicators, Eq. 7 calculates the maximum value, which is designated as the GWF index (Hoekstra et al. 2011). This index serves as a comprehensive metric, encapsulating the intricate relationship between the dilution processes and pollutant concentrations, providing a quantitative measure of the environmental repercussions associated with various pollution indicators.

$$\mathrm{GWF} = \mathrm{Max}\ \{\mathrm{GWF_1}, \mathrm{GWF_2}, \mathrm{GWF_3}, \ldots, \mathrm{GWF_i}\} \tag{7}$$

Industries are increasingly aware of their contribution to water scarcity and pollution, necessitating a response. The industrial WF methodology specifically examines the production, providing a visual representation of freshwater consumption and environmental impacts. Notably, case studies highlight the significant role of industrial GWF, which quantifies the volume of freshwater required to dilute toxic pollutants generated by industrial enterprises to meet the water quality standards. For example, chemicals present in fabrics are washed out by water during the manufacturing process. Consequently, the resulting effluent water becomes contaminated with these chemicals, contributing to the GWF. Assessing the GWF per unit of production allows us to consider the environmental impact of industrial enterprises. The effluent water generated in the wet processing unit is quantified by multiplying the total weight of the finished product with the water consumption specific to each process. This effluent water contributes to the GWF of the wet processing units. For instance, in the production of woven and knitted fabrics, the average GWF is 16,000 and 3300 L/kg of product, respectively (Kumar et al. 2021).

The WF of production ($\mathrm{WF_P}$) quantifies the freshwater consumption required for assimilating pollutants and the production output within the Industrial Park. To calculate $\mathrm{WF_P}$, the sum of the net blue WF (BWF), which is consumed in industries, and grey WF (GWF) is divided by the total production yield (Y), using Eq. 8 (Hoekstra et al. 2011):

$$\mathrm{WF_P} = \frac{\mathrm{BWF} + \mathrm{GWF}}{Y} \tag{8}$$

The GWF is a crucial metric for assessing the environmental impact of pollutants released during industrial production. Therefore, the evaluation of embedded water required for pollution dilution per ton of production ($\mathrm{GWF_P}$) is calculated using Eq. 9 (Hoekstra et al. 2011):

$$\mathrm{GWF_P} = \frac{\mathrm{GWF}}{Y} \tag{9}$$

These basic calculations must be computed with more detail under different scenarios, especially for industrial wastewater.

2.2.2 Multiple Outlets in the Same Catchment

GWF assesses the environmental impact of water consumption due to assimilation, particularly in the context of diverse discharge sources (represented by k), as shown in Fig. 5. This condition is practical when different industrial units (or various industrial treatment modules) discharge their effluents into different locations in the same catchment.

Therefore, GWF in the receiving water bodies is affected by multiple outlets that should be considered as introduced in this study by Eq. 10 (by authors). The pollution load of each outlet ($Q_k \times C_k$) is identified and computed for a specific pollutant (GWF_i) by evaluating the cumulative effect of these discharge sources, taking into account both the quality and consumption aspects. It is important to note that if the GWF value is obtained negative in one outlet, it will appear with the same negative sign in the equation and will not be considered zero. This is because a negative footprint reflects the extent of diluting the water body due to the cleaner pollution level of effluent compared to the environment, rather than the volume of discharged effluent dilution. Therefore, in the final aggregation of discharged pollution load, both positive and negative contributions should be considered together to determine the ultimate impact of pollution on the specific water body.

$$\mathrm{GWF_i} = \left(\frac{\left(\sum_{k=1}^{n} Q_k \times C_k \right) - \left(Q_{abstr} \times C_{act} \right)}{C_{max} - C_{nat}} \right)_i \qquad (10)$$

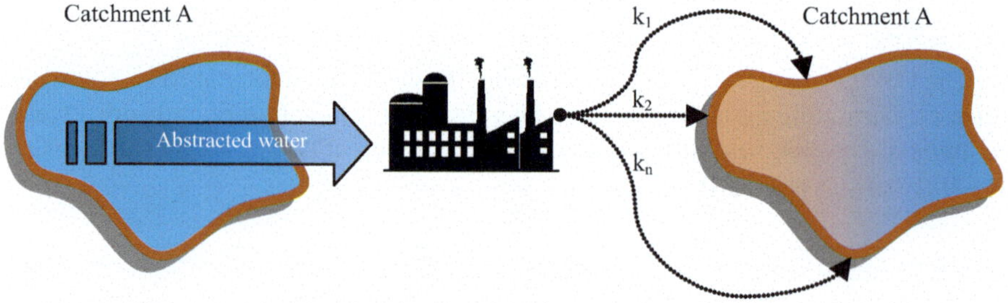

Fig. 5 GWF assessment where the effluents are spatially discharged through multiple wastewater treatment units. (By authors)

2.2.3 Abstracted Water from Different Sources

The evaluation of environmental impacts of pollutants (i) in discharged industrial wastewater involves considering the withdrawal of water from various catchments. These catchments exhibit different C_{act} values, which directly influence the GWF. The interplay between these factors is depicted in Fig. 6, providing insights into the complex relationship between pollution, water withdrawal, and environmental consequences.

The term intake water concentration calculates a weighted mean of existing pollution in the inlet, factoring in the influence of each pollutant source based on its flow rate as presented in this study by Eq. 11 (by authors). This obviously recognizes that higher flow rates among income sources contribute more significantly to the amount of net GWF. By integrating these aspects, the equation provides a robust method for quantifying the GWF, aiding industries in sustainable water management and environmental compliance by understanding pollutant dynamics and their impact on water quality.

$$\mathrm{GWF_i} = \left(Q_{\mathrm{k}} \times \left[\frac{C_{\mathrm{k}} - \left(\frac{\sum_{j=1}^{n} C_{\mathrm{j}} \times Q_{\mathrm{j}}}{\sum_{j=1}^{n} Q_{\mathrm{j}}} \right)}{C_{\mathrm{max}} - C_{\mathrm{nat}}} \right] \right)_{\mathrm{i}} \tag{11}$$

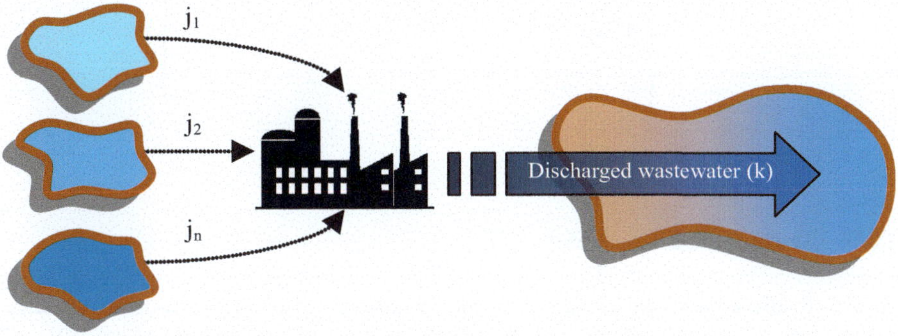

Fig. 6 GWF assessment during abstracted water from different sources (by authors)

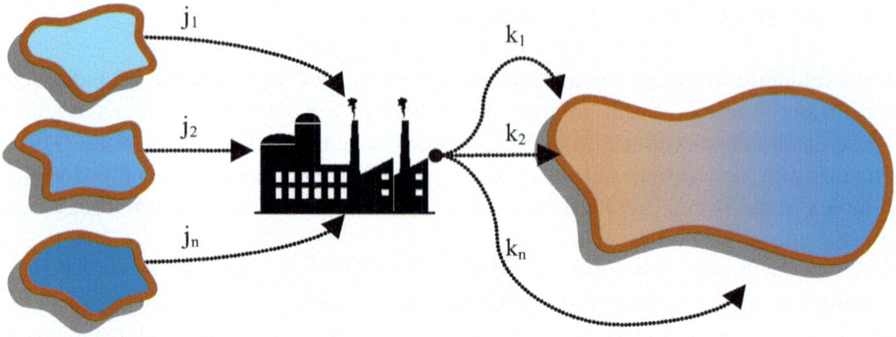

Fig. 7 Abstracted water from different sources (j) and discharged effluents (k) by multiple industrial processes or treatment units (by authors)

2.2.4 Different Sources and Outlets

Figure 7 shows that both abstraction and discharge involve multiple sources; the equation for the GWF encapsulates a comprehensive assessment of pollution in industrial wastewater processes.

The subsequent expression involves the weighted difference in pollutant concentrations between water abstraction and wastewater discharge, accounting for the proportional influence of each source based on its flow rate, as proposed in this study by Eq. 12 (by authors). This weighted arithmetic mean for C_{eff} and C_{act} emphasizes the collective impact on water quality. The denominator, comprising the permissible pollutant levels and baseline water quality, contextualizes the environmental standards. This equation provides a framework for evaluating the GWF, considering the complexity of multiple sources in both abstraction and discharge, offering insights into the environmental ramifications of industrial activities related to pollution.

$$\text{GWF}_{\text{i}} = \left(\left(\sum_{k=1}^{n} \text{Q}_k \right) \times \left[\frac{\left(\frac{\sum_{k=1}^{n} C_{\text{k}} \times Q_{\text{k}}}{\sum_{k=1}^{n} Q_{\text{k}}} \right) - \left(\frac{\sum_{j=1}^{n} C_{\text{j}} \times Q_{\text{j}}}{\sum_{j=1}^{n} Q_{\text{j}}} \right)}{C_{\max} - C_{\text{nat}}} \right] \right)_{\text{i}} \tag{12}$$

2.2.5 Reuse Reclaimed Water

Figure 8 illustrates the dynamic of an industry with a closed-loop system, underscoring the pivotal role of "Reuse reclaimed water" in reducing the GWF.

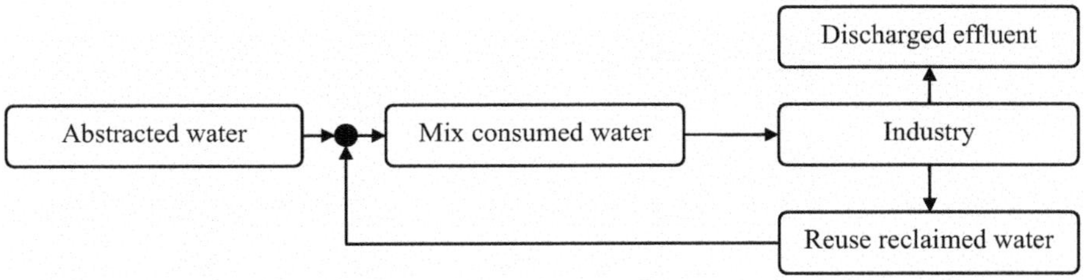

Fig. 8 The concept of reuse reclaimed water as an industrial water resource (by authors)

When an industrial outlet, instead of being discharged, is reintegrated as the mix of consumed water (as shown in Fig. 8), it can significantly curtail water consumption in the industry. This dual benefit alleviates strain on the external catchment. In the absence of effluent reuse, the GWF would surge, paralleled by an increased demand for water. This closed-loop strategy not only ensures regulatory compliance but also epitomizes a sustainable approach. By mitigating water consumption and easing pressure on water bodies (BWF), the incorporation of "reuse reclaimed water" emerges as a cornerstone in promoting resource efficiency and environmental responsibility within industrial water management. Reusing industrial effluents affects the GWF. The mixed wastewater concentration, which acts as the new actual concentration (C'_{act}), should be substituted in place of the prior C_{act} ($C_{act} = C'_{act}$) and is determined as introduced in this study by Eq. 13 (by authors). In this equation, except for C_{act}, the determination of the final GWF is intricately tied to the dynamic quantifications of reuse reclaimed water concentration (C_{re}) and its flow rate (Q_{re}), whereas both factors can be influenced by the subject and limitations of reused wastewater. In this case, and due to the complexity and variations of C_{re}, GWF can be just based on intake water concentration considered by Eq. 5. For defining the C_{max} of pollutants, it is essential to consider the standard for reuse and recycling effluents instead of the wastewater discharge standard. However, the high sensitivity of the freshwater recycling standard leads to the overestimation of the WF. Accordingly, we strongly recommend that GWF reduction by reusing treated industrial wastewater should be studied further in future case studies.

$$C'_{act} = \left(\frac{(Q_{re} \times C_{re}) + (Q_{abstr} \times C_{act})}{Q_{re} + Q_{abstr}} \right)_i \quad (13)$$

2.2.6 Conditions and Scenarios

GWF is calculated using ambient water quality standards, which vary due to the location and pollutant type. When local standards are unavailable, the strictest water quality standards (C_{max}), e.g., EU, US-EPA, or CCME, are recommended. For transboundary studies, it is recommended to use a consistent set of standards to

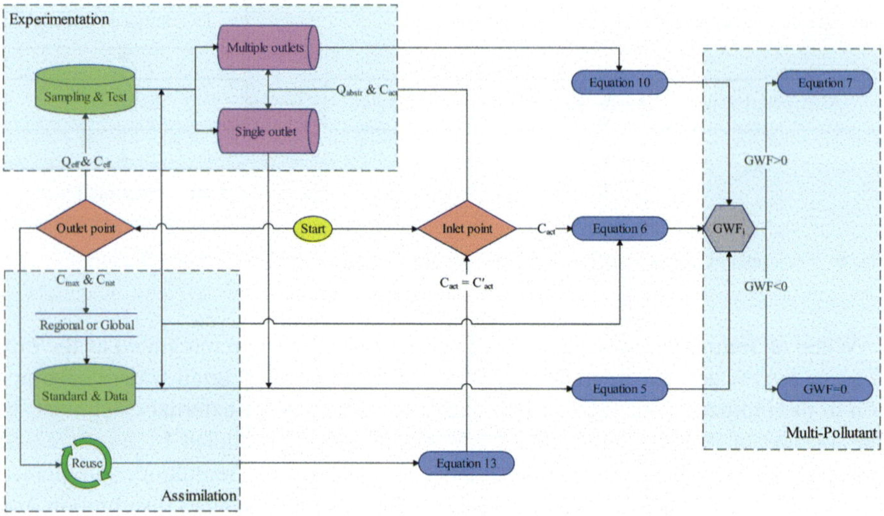

Fig. 9 Flowchart of GWF assessment by equations and variables (by authors)

prevent discrepancies due to varying national legislations (Franke et al. 2013). In GWF assessments, it is essential to specify the water quality standards used to maintain transparency. When discharging pollutants into various targets, exploring different scenarios helps determine the necessary dilution volumes. Due to regional variations in natural concentrations influenced by geology and climate, using local data is preferable over a global standard. If local data is lacking, established guidelines can provide substitute values. This approach ensures accurate GWF calculations and considers the environmental impact of industrial activities on water bodies. Accounting for the presence or absence of pollutant concentrations in C_{act} is a critical variable influencing the GWF. Often, studies may neglect the significance of these concentrations in intake water. Yet, for an accurate assessment of industrial GWF, it is imperative to acknowledge the impact of antecedent pollutants unrelated to the industrial entity. These prior contaminants must be excluded from the total pollution generated to ascertain the genuine GWF.

GWF evaluates the environmental impact of pollution points associated with water extraction from one location and the discharge of effluents at the same catchment (Eq. 5) or another destination (Eq. 6), as shown in Fig. 9 (Hoekstra et al. 2011). Consideration of the flow rate and the amount of pollutants, especially in industrial wastewater with a high range of variations, is essential. In the case of industrial effluents, calculating pollutant loads involves measuring both flow rates and pollutant concentrations. Importantly, the term "pollution load" specifically denotes the additional amount introduced into the receiving environment, distinguishing it from the pollution load already present in the water body (Franke et al. 2013). Observations indicate that, in many instances, pollution loads discharged into the receiving environment are either equal to or greater than the existing pollution loads in the environment. However, there are certain conditions for determining the GWF index. In the context of GWF classification, if the C_{eff} equals C_{act}, there is no impact (GWF = 0) as the water quality remains unchanged. When C_{eff} equals C_{max}, the GWF is positive (GWF > 0), indicating an increase in water quality. A similar positive GWF results when C_{act} aligns with C_{nat}, suggesting the water closely resembles its natural state. Conversely, if C_{eff} is less than C_{act}, the GWF is negative (GWF < 0), reflecting higher wastewater quality than consumed or environmental water. A negative GWF also occurs when C_{max} is less than C_{nat}, indicating a cleaner discharge into the environment. Lastly, if C_{max} equals C_{nat}, the GWF is considered to have an infinite impact, especially under minimal pollution conditions, although this scenario is not typically regulated by water and wastewater quality standards. These scenarios collectively offer a comprehensive understanding of GWF's influence on water quality. Figure 9 shows the complete path of calculating the GWF index based on the variables of this index and the components introduced in the previous sections. Furthermore, this figure is prepared for a part of the proposed equations in this study (Eq. 10). It is possible to provide a similar flowchart for the remaining equations (e.g., Eqs. 11 and 12).

3 Study Area: Industrial Park

3.1 *Introduction*

Isfahan City, in Iran, experiences a semi-arid climate characterized by diverse temperature extremes annually. Table 8 shows that the annual average temperature is 16.7 °C. Despite an annual precipitation level of 111.5 mm for a period of about 44 days, the area is notably dry, with an average relative humidity of 29%. Sunshine is abundant, with 3280 h annually and the average wind speed is 3 m/s (Hasehzadeh

Table 8 Regional climate profile and weather factors (Hasehzadeh et al. 2018; Yazdani 2019)

Climatic parameters	Unit	Value
Annual average temperature	°C	16.7
Annual average minimum temperature	°C	−1
Annual average maximum temperature	°C	38.3
Annual average relative humidity	%	29
Annual average wind speed	m/s	3
Number of rainy days	day	44
Annual average snowy days	day	7
Annual average frost days	day	69
Annual sunshine hours	h	3280
Annual average precipitation	mm	112
Maximum daily precipitation	mm	48
Average daily evapotranspiration	mm	77

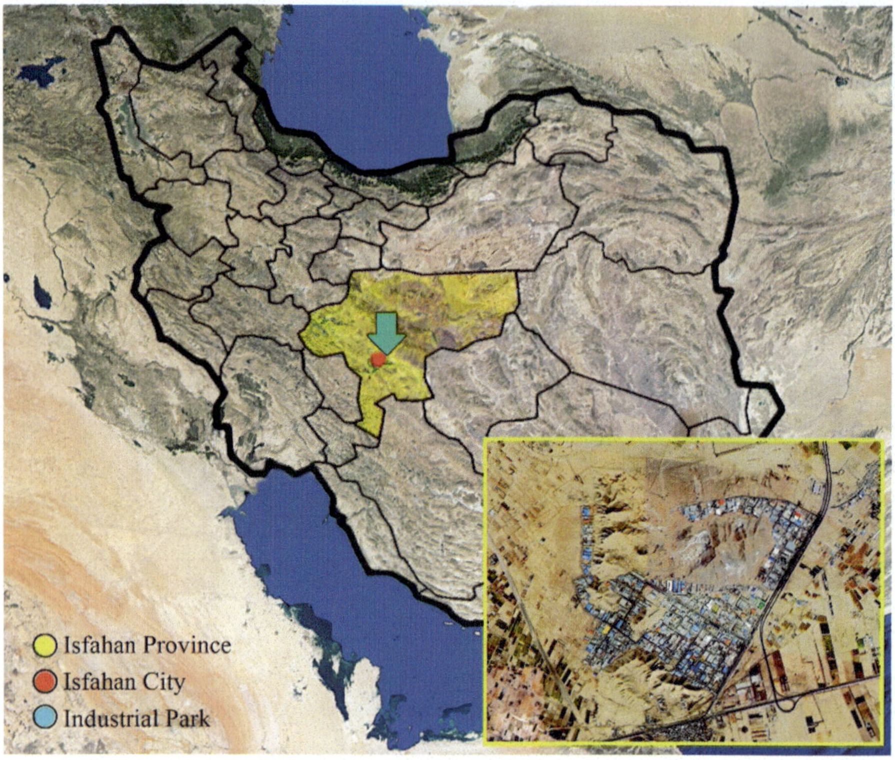

Fig. 10 Location of the Industrial Park in Isfahan Province (by authors)

et al. 2018; Yazdani 2019). In general, Isfahan's climate is characterized by significant temperature variations, limited precipitation, and ample sunshine.

Figure 10 illustrates the Industrial Park considered for this study, which is located in Isfahan Province, Central Iran. This zone has latitude and longitude of 32°40′ 2.74″ N and 51°50′52.68″ E, respectively with an approximate area of 3.1 square kilometers. The Industrial Park stands as the largest industrial city in the Isfahan Province (Iran).

This area is composed of several industrial, manufacturing, and trade units. The total number of active units in this industrial zone is 554 enterprises and sectors. It is important to note that some active industrial units in this park, depending on their type of activity and the quantity of wastewater production, have pre-treatment units. These pre-treatment units ensure that the effluents reach the permissible level before discharging into the collection system for centralized treatment. The final destination of the largest portion of this complex industrial effluent is the irrigation of non-fruit-bearing trees. The industrial activities in this area are different. Based on available data on 2024, the Industrial Park studied mainly includes activities like metal and plating industries (24.4%), textile and dyeing industries (21%) as shown in Fig. 11, chemical and toxic industries (18.4%), food and dairy industries (14.2%), cellulose

Fig. 11 Industrial units of textile and dyeing sectors (as an example) in the Industrial Park (by authors)

Table 9 Methods for conducting tests for conventional and non-conventional pollutants (APHA 2012; ASTM 2023)

Pollutants	Unit	Test method
Biochemical oxidation demand (BOD)	mg/L	Standard 5210 D
Chemical oxidation demand (COD)	mg/L	Standard 5220 D
Phosphorus (P), Phosphates (PO_4)	mg/L	Standard 4500 P D
Ammonia (NH_3)	mg/L	Standard 4500 NH_3^+ B & C
Ammonium (NH_4)	mg/L	Standard 4500 NH_4^+ B & C
Nitrite (NO_2)	mg/L	Standard 4500 NO_2^- B
Nitrate (NO_3)	mg/L	Standard 4500 NO_3^- B
Chromium (Cr), Copper (Cu), Lead (Pb)	μg/L	ASTM E2823–17
Zinc (Zn)	μg/L	ASTM D5673–15

and paper industries (6%), electrical and electronic industries (6%), non-metallic mining industries (3%), and the remaining portion is attributed to service companies, administrative offices, and small industries like packaging.

3.2 *Sampling and Experimentations*

The examination of multiple pollutants in industrial untreated wastewater involves analyzing various parameters, such as conventional indicators and metals as depicted in Table 9 (APHA 2012; ASTM 2023).

These pollutants are typically analyzed in the laboratory following the standard methods for the examination of water and wastewater (APHA) and the American Society for Testing and Materials (ASTM) to ensure accurate and standardized results (APHA 2012; ASTM 2023). The evaluation of HMs (chromium, copper, lead, and zinc) was conducted using inductively coupled plasma mass spectrometry (ICP-MS). This technique allows for more precise measurement (parts per billion) of

HMs in the Industrial Park effluents, ensuring compliance with environmental regulations and calculation of the GWF.

3.3 Blue Water Footprint (BWF)

In 2023, wastewater flow rates in the Industrial Park exhibited seasonal variations (Fig. 12). The annual average flow rate was approximately 1057 m^3/day. During October, November, and December, the average flow rate was 1146 m^3/day, with a notable peak at 2717 m^3/day (November) and a minimum of 415 m^3/day (October). In January, February, and March, the average flow rate increased to 1719 m^3/day, with 50% rise from autumn. The highest and lowest flow rates were 3337 m^3/day (March) and 734 m^3/day (February), respectively. These variations emphasize the characteristics of industrial wastewater generation, highlighting the need for nuanced assessments of the GWF.

Figure 13 shows the daily average withdrawal of water which has entered the Industrial Park, and then the proportions (%) of the used water (as BWF) and produced wastewater. In October, the abstracted water flow rate was 2929 m^3/day, with 34.5% (1008 m^3/day) of wastewater generation ratio. Daily withdrawal of water in November and December was determined to be about 2990 m^3/day (44% wastewater production) and 2862 m^3/day (42% wastewater production), respectively. In the next 3 months, the abstracted water in the Industrial Park considerably increased from 3045 m^3/day in January to 3573 m^3/day in March. The ratios of industrial wastewater production in winter assessed in January, February, and March were around 47%, 51%, and 57% of water withdrawal, respectively. Notably, the average volumes of abstracted water and produced wastewater in this period were calculated at about 3118 ± 111 m^3/day and 1450 ± 151 m^3/day, respectively. The magnitude of water consumption in the Industrial Park during manufacturing processes such as food production (as BWF) in the studied period counted about 1921 m^3/day in October (maximum value), whereas in March it was equal to 1533 m^3/day (minimum). The mean value of net BWF in industrial enterprises was determined to be approximately 1668 ± 54 m^3/day.

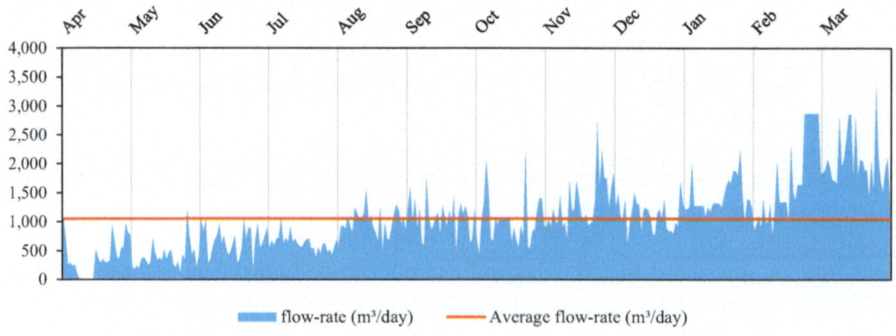

Fig. 12 Daily flow rate of wastewater discharged from the Industrial Park (by authors)

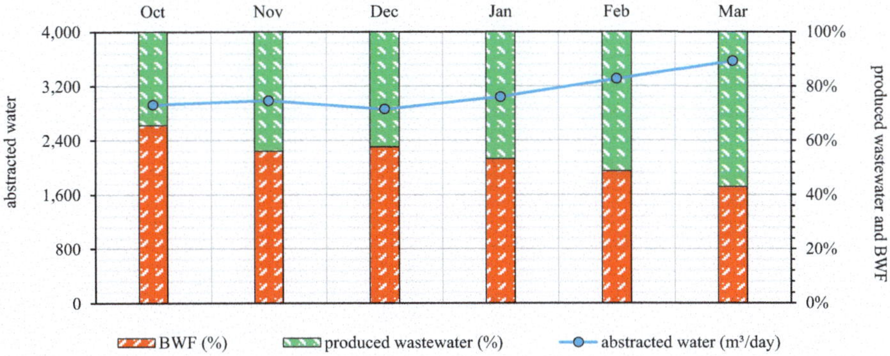

Fig. 13 Water withdrawal, generated wastewater, and BWF in the Industrial Park (by authors)

In addition to the variables mentioned earlier, the GWF also reflects the intricate interplay between water withdrawal and wastewater production. Notably, seasonal variations affect this delicate balance. In winter, when water consumption increases due to manufacturing processes (e.g. food production and textile), GWF has higher fluctuations. These dynamics underscore the need for sustainable water management practices within industrial parks, emphasizing efficient water use.

3.4 Water Quality Assessment

The extensive dataset reveals a detailed view of pollutant concentrations ($C_{\text{eff-i}}$) for the "i" parameters at the Industrial Park over a period of six months, as shown in Fig. 14. The BOD parameter at the Industrial Park traces a narrative of change, starting at 1020 mg/L in October, reaching a low concentration of 250 mg/L in January, and then rising dramatically to the maximum concentration of 2450 mg/L in March. In parallel, the COD parameter reveals a tale of escalation, increasing from 1076 mg/L in October to its peak of 2845 mg/L in March. Meanwhile, the *P* parameter fluctuates, with a maximum concentration of 8.4 mg/L in November. These variations verify the gradually increasing complexity of industrial wastewater due to the diverse range of production processes. The highest monthly concentration of PO_4 was 196 mg/L in November. The maximum concentration of NH_3 was 108 mg/L, in March. For NH_4, the maximum concentration assessed was about 114 mg/L. However, the maximum concentrations of NO_2 and NO_3 were 0.21 and 20.8 mg/L, respectively. Regarding the HMs that were examined in the industrial wastewater, the highest concentration was about 164.08 μg/L for Zn. However, Cr, Cu, and Pb had lower concentrations compared with Zn. Furthermore, the data highlights the critical need for sustainable practices and stringent environmental regulations. As the BOD and COD concentrations fluctuate, industrial GWF underscores the importance of continuous monitoring and effective treatment methods for multiple pollutants.

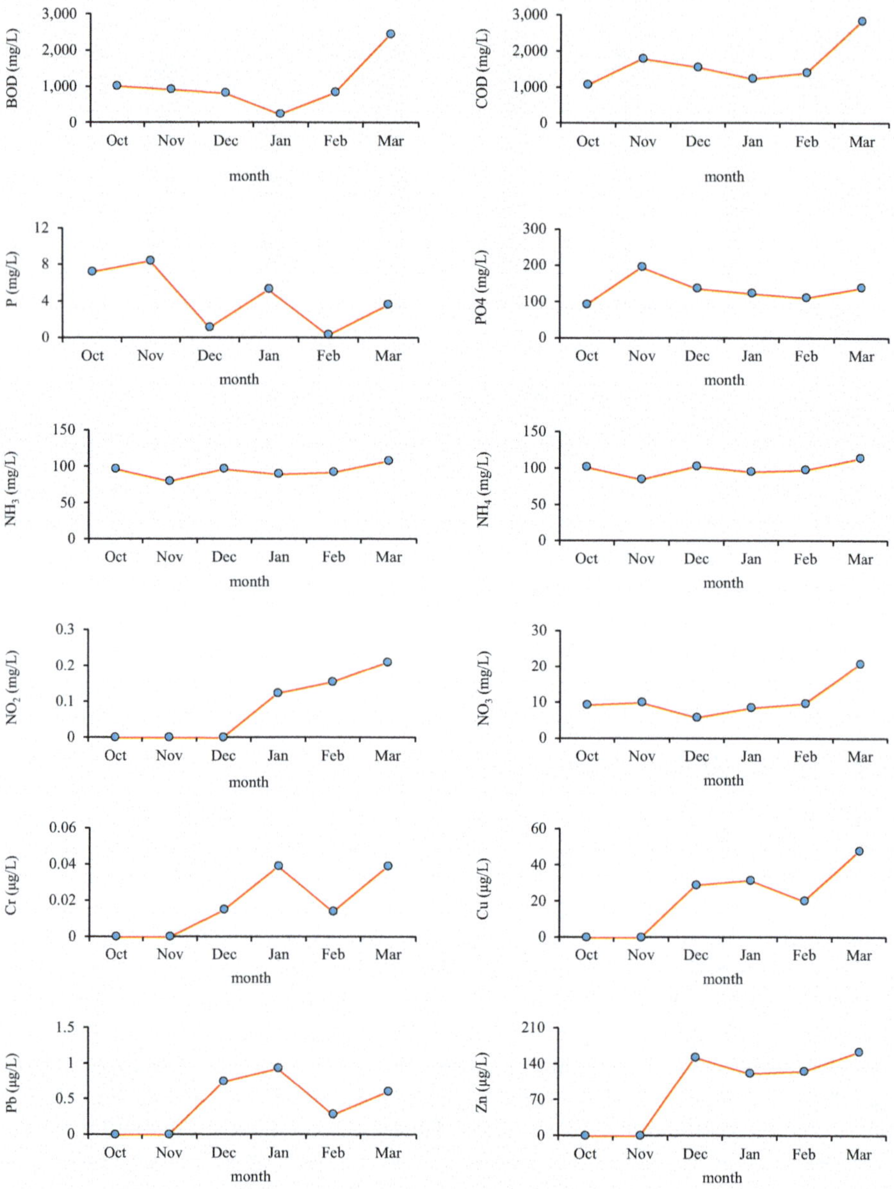

Fig. 14 Monthly concentrations of the pollutants examined ($C_{eff\text{-}i}$) in the Industrial Park wastewater (by authors)

The variations of HM concentrations, even when detected at levels close to zero (October and November), necessitate further investigations about their potential adverse impacts on the ecosystems.

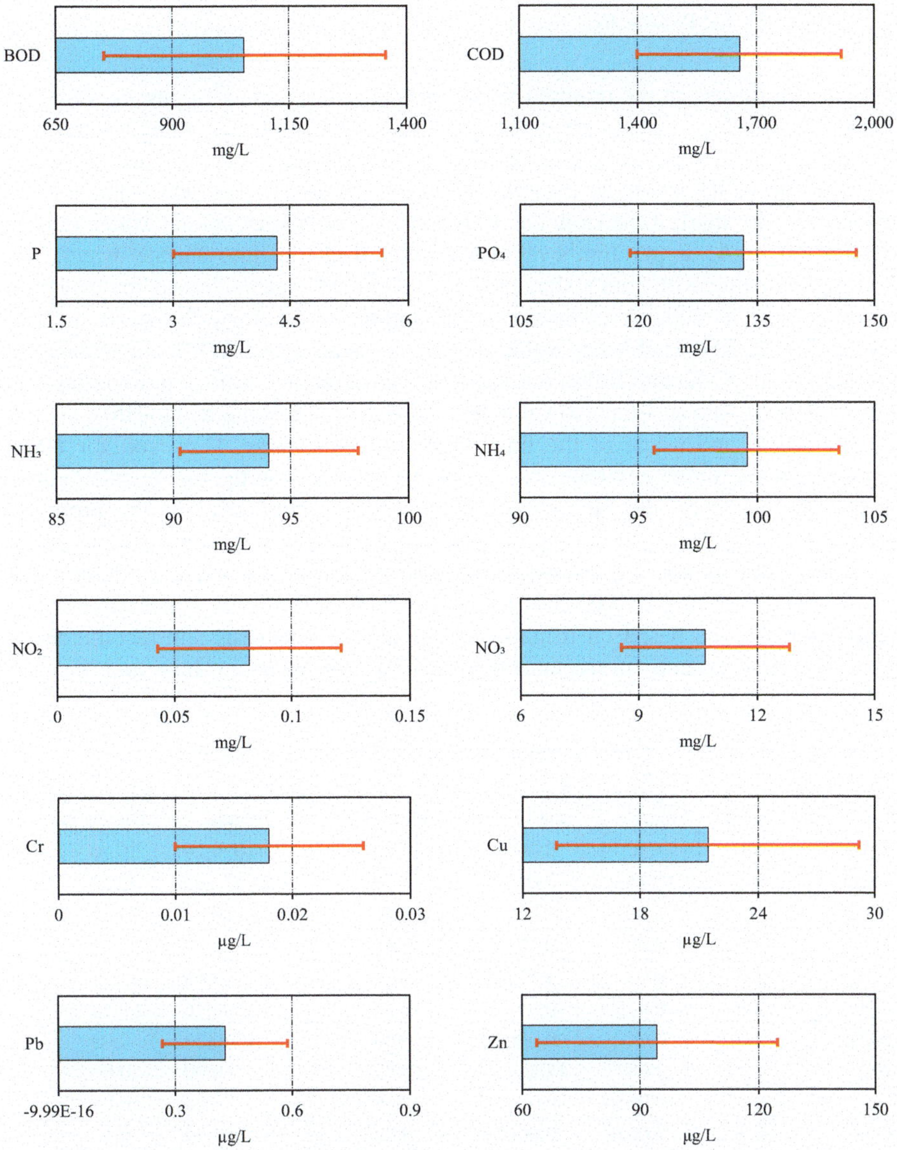

Fig. 15 Monthly average of concentration (C_i) with standard error in the Industrial Park wastewater (by authors)

Furthermore, for a better appraisal of pollutant concentrations in the wastewater, the average values of these parameters (C_i) have been specified with standard error in Fig. 15. However, the average pollutant concentrations have not been used in determining the GWF index, as this index is calculated based on monthly assessments and then averaged.

3.5 GWF Assessment

The concentrations of the required variables (C_{max}, C_{nat}, and C_{act}) are presented in Table 10. The C_{max} values are determined by guidelines from different countries, based on various regulatory standards for each parameter. For instance, the C_{max} variable for COD is set at 7 mg/L following Moldova's surface water quality guidelines, 30 mg/L based on the European Union's recreational water quality criteria, 60 mg/L as per Iran's regulations for discharge into surface water, and 200 mg/L in line with Iran's standards for reclaimed water released for irrigation purposes (OECD 2006; Vice President's Planning and Strategic Supervision 2010; OECD 2007). The established value of C_{max} to determine GWF for conventional pollutants (e.g. COD and BOD) is considered based on the third scenario outlined in Table 7, taking into account a lower level of pollution risk compared to HMs, as well as destination and usage of the final wastewater discharge. However, for HMs, considering the higher environmental risks, their impacts on the biological environment, and based on the effluent destination, the second level of the proposed scenarios (Table 7) has been employed with a greater level of stringency to determine the GWF index. For example, the selected values for Cr, Cu, Pb, and Zn, respectively, were set at 100, 200, 1000, and 2000 μg/L. These concentrations for HMs were derived based on irrigation water quality standards and recommended limits for constituents in reclaimed water for irrigation in Iran and the US

Table 10 C_{max}, C_{nat}, C_{act}, and C, monthly average of GWF, and D_f (by authors)

Pollutants	C_{max}[1] mg/L	C_{nat}[2] mg/L	C_{act}[3] mg/L	C_i mg/L	GWF m^3/month	D_f Dimensionless
BOD	30	1.50	3.00	1054.67	1,710,029.29	36.90
COD	60	10.15	2.00	1658.50	1,499,421.69	33.23
P	0.4	0.002	0.00	4.34	437,660.84	10.88
PO_4	6.0	0.09	0.10	133.34	965,837.79	22.54
NH_3	2.0	0.05	1.50	94.09	2,002,963.09	47.47
NH_4	2.0	0.015	0.05	99.59	2,155,202.16	50.14
NO_2	3.0	0.04	0.006	0.08	1340.73	0.03
NO_3	11.3	2.08	17.2	10.69	0.00	0.00
Cr	0.1	0.0036	0.00	0.00002	9.04	<0.01
Cu	0.2	0.0014	0.00075	0.022	4954.79	0.1
Pb	1.0	0.0018	0.00	0.00043	19.19	<0.01
Zn	2.0	0.057	0.085	0.095	77.15	<0.01

References C_{max}[1]: (Department of Environment Islamic Republic of Iran 2016; OECD 2006; Department of Environment Islamic Republic of Iran 2023; Vice President's Planning and Strategic Supervision 2010; Health Ministry of New zealand Government 2019; Fipps 2003; Australian and New Zealand Council 2000)

References C_{nat}[2]: (Franke et al. 2013; UN Environment Programme GEMStat 1980a, b; Environmental Protection Organization of Iran 1979; Tedd et al. 2017)

References C_{act}[3]: (Department of Environment Islamic Republic of Iran 2016; Institute of Standards and Industrial Research of Iran 2020; Arastou 2024)

(e.g. 0.1 mg/L for Cr) (Vice President's Planning and Strategic Supervision 2010; Fipps 2003).

Another variable, C_{nat}, was derived from the Global Freshwater Quality Database (GEMStat). Its information was provided based on scientifically sound data on the state and trend of Iran's water quality (ZayandehRoud in Isfahan) in 1982 (UN Environment Programme GEMStat 1980a; UN Environment Programme GEMStat 1980b). Moreover, the selected concentrations were also checked with local data provided in 1979 for parameters such as COD, PO_4, NO_2, Cr, Pb, and Zn (Environmental Protection Organization of Iran 1979). Owing to the limited data to achieve C_{nat} for NH_4 and Cu, these two values were estimated based on the global numbers and reliable references (Franke et al. 2013).

C_{act} for some pollutants (e.g. P, Cr, and Pb) in freshwater was determined as zero. The highest value for C_{act}, according to the Industrial Park's reports, is set at 17.2 mg/L for NO_3. Then, the maximum D_f was 50.14 times for NH_4 in the Park (Table 10). Notably, the D_f was found to be negative ($D_f < 0$) for NO_3 and assumed to be zero. This negative value for D_f indicates that the wastewater, based on NO_3 in the Industrial Park, is cleaner than the consumed water ($C_{eff} < C_{act}$). Consequently, the water body is not acting as a diluent for the wastewater; instead, it is being further diluted.

Our findings indicate that GWF should be determined by examining the influence of each parameter while simultaneously accounting for the combined effects of a range of pollutants for a more accurate evaluation. It worths mentioning that by an average of generated industrial effluent (1450 m^3/day), the GWF based on the D_f is estimated about 2,181,209 (NH_4) m^3/month.

The monthly average GWF index of the Industrial Park is given in Table 10 and determined as the mean of monthly GWF values. This index for conventional pollutants was between 0 (NO_3) and 2.155 (NH_4) MCM/month, whereas the maximum GWF based on non-conventional pollutants (HM_s) was evaluated about 0.005 MCM/month for Cu. Figure 16 provides an accurate overview of the average GWF with the standard errors of the wastewater for the studied Industrial Park. Here, for example, the BOD- and COD-related GWFs are 1.711 $\pm$ 0.691 MCM/month and 1.5 $\pm$ 0.393 MCM/month, respectively. However, the maximum GWF is attributed to NH_4 as it is equal to 2.155 $\pm$ 0.285 MCM/month while NH_3- and PO_4-related GWFs, with values of 2.0 $\pm$ 0.273 MCM/month and 0.966 $\pm$ 0.139 MCM/month, are next in order, respectively. These GWFs show the importance of nutrients in industrial wastewater and their further ecological impacts on the environment. Nutrients in industrial wastewater can significantly affect ecosystems, leading to some critical issues such as eutrophication (Billen et al. 1999), which can disrupt aquatic life by promoting excessive plant growth and depleting oxygen in water bodies (Chislock et al. 2013). The GWF for PO_4 was determined to be 0.966 $\pm$ 0.139 MCM/month. We should also note that the GWF of NO_2, often a byproduct of industrial processes, is 1340 $\pm$ 676 m^3/month, while the GWF of NO_3 diminished to zero (GWF $<$ 0), owing to the greater pollution load in the abstracted water compared to the wastewater ($C_{eff} < C_{act}$) and the real assimilation effects of flow rate.

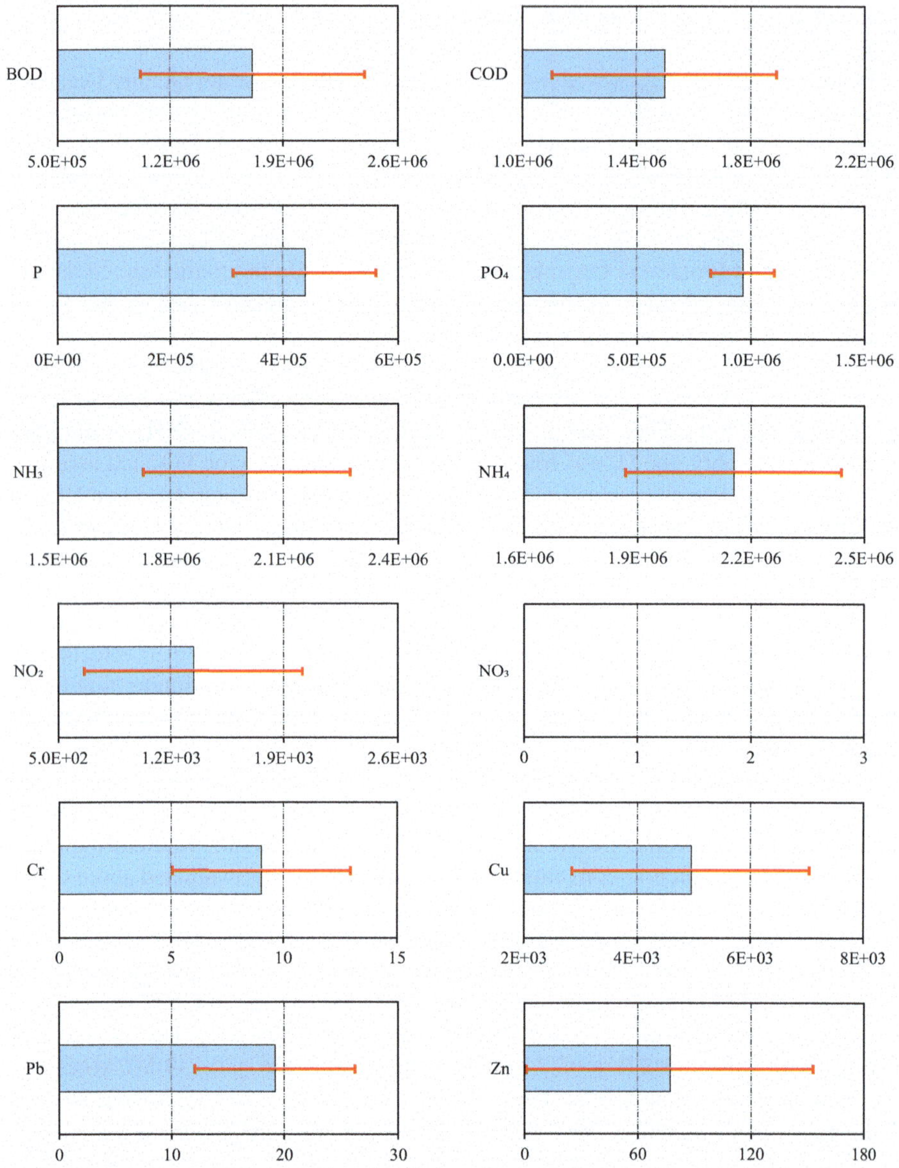

Fig. 16 Monthly average of GWF with standard errorr (m^3/month) in the Industrial Park wastewater (by authors)

The GWF for Cr, a metal that can be toxic in high concentrations, is 9 ± 3.95 m^3/month, and it is closely monitored to prevent ecological harm. Cu metal, essential in small amounts but harmful in excess, has a footprint of 4955 ± 2096 m^3/month. Pb, known for its potential toxicity, has a footprint of 19 ± 7 m^3/month. Lastly, Zn, vital

Fig. 17 D_f index (dimensionless) for assimilation of pollution (by authors)

for life and detrimental in higher concentrations, has a footprint of 77 ± 77 m^3/month, reflecting the approach to its management. Despite the toxicity of metal compounds, the environmental impact of the entire Industrial Park was assessed based on the conventional parameters of industrial wastewater, with a focus on the level of NH_4.

Furthermore, the amount of WF is calculated concerning the production yield and the total embedded water. The WF_P measured the total freshwater used to dilute pollutants and for production in the Industrial Park. It is calculated by adding the net BWF, "water used in industries", and the net GWF, "required water for pollution assimilation by production", and then dividing by the total yield (Y), as shown in Eq. 8. The total amount of production in the Industrial Park (total yield) is estimated at around 1.257 ± 0.1 million-ton/month. Accordingly, the GWF per ton of production (GWF_P), which assessed the environmental impact of pollution, is calculated by dividing the GWF by total yield, as in Eq. 9. The values of average WF_P and GWF_P, based on the mean value of net BWF (1668 m^3/day) and GWF (71,833 m^3/day by NH_4) in the Industrial Park, are estimated at 1.76 and 1.72 m^3/ton, respectively. It shows the amount of water consumed (as BWF added to GWF) per ton of production in the Industrial Park.

Based on the results, the D_f (equal to the maximum amount) for the receiving water body was about 50 under the NH_4 parameter, as shown in Fig. 17. It means that the volume of polluted effluent, including pollutants like BOD, COD, P, PO_4, NH_3, NH_4, and other parameters must be diluted with the ambient water approximately 50 times to reduce the pollution concentration to the allowable limits. This dilution virtually ensures that the concentrations of all pollutants in the effluent are decreased to a safe level for the environment and it now complies with water quality standards. The D_f helps to measure the impact of industrial discharge on water bodies and the extent of treatment required before releasing the wastewater into the environment.

3.6 WF by Production Assessment

The total WF for the Industrial Park has been estimated based on the pollutants studied according to the WF, GWF, WF_P, and GWF_P. Table 11 outlined the WF and GWF for the Industrial Park over a period of six months, with a focus on the leading pollutants for each month. In October, the WF was 1.609 MCM/month, and the

Table 11 Monthly average WF_P and GWF_P of the Industrial Park (by authors)

Month	Leading pollutant	WF	GWF	WF_P	GWF_P	GWF/BWF
		m^3/month		m^3/ton		Dimensionless
Oct	NH_4	1,609,311	1,551,681	1.29	1.24	26/93
Nov	NH_4	1,738,155	1,687,935	1.39	1.35	33/62
Dec	NH_4	1,779,108	1,729,638	1.42	1.38	34/97
Jan	NH_4	2,104,450	2,055,880	1.68	1.64	42/33
Feb	NH_4	2,559,507	2,511,087	2.04	1.99	51/87
Mar	BOD	5,119,234	5,074,777	4.08	4.04	114/16

GWF was 1.552 MCM/month, with NH_4 as the leading pollutant. November saw a slight increase, with the WF at 1.738 MCM/month and the GWF at 1.688 MCM/month. December's figures were close, with a WF of 1.779 MCM/month and a GWF of 1.729 MCM/month. A more substantial rise occurred in January, where the WF reached 2.104 MCM/month and the GWF was 2.056 MCM/month. February's numbers climbed higher, with a WF of 2.559 MCM/month and a GWF of 2.511 MCM/month. However, the most significant jump was in March, when the leading pollutant changed to the BOD parameter, and the WF skyrocketed to 5.119 MCM/month (5.075 GWF and 0.045 BWF), alongside a GWF of 5.075 MCM/month. This data indicated a progressive increase in water consumption and pollution levels, particularly highlighted by the sharp increase in March, which was attributed to specific industrial activities or changes in the park's operations in this month.

To assess the proportion of GWF pressure relative to BWF, Table 11 presents the GWF/BWF ratio. This ratio quantifies the GWF pressure on the catchment per unit of industrial water consumption, measured as BWF in m^3/month. The peak value of this ratio was recorded in March at 114.2, indicating that for every cubic meter of water used by the Industrial Park monthly, approximately 114.2 m^3/month is required by water bodies to dilute pollution from BOD. The average GWF/BWF ratio during the study period was approximately 50.7 $\pm$ 13.2.

The WF_P refers to the amount of water (m^3) required to assimilate pollution and the volume of water consumed in the industry per unit of production (ton). Provided that this index is calculated solely based on dilution load, it represents the GWF_P. This number evaluates insights into the needed freshwater to mitigate pollution effects. Implementing sustainable practices and minimizing WF_P and GWF_P are essential for responsible industrial development. Monthly WF_P is computed based on data from Table 11. Notably, in the Industrial Park, the highest WF_P by production (about 1.257 million tons per month) occurred in March at 4.08 m^3/ton. The table displays a monthly breakdown of WFs, represented as ratios of consumed water to production, denoted by WF_P and GWF_P. These ratios provide a comparative analysis of water usage efficiency within an industrial setting. The data shows a consistent pattern of increasing ratios from October to March, indicating a growing demand for water relative to production output. In October, the WF_P ratio was 1.29 m^3/ton, and the GWF_P ratio was about 1.24 m^3/ton, suggesting lesser water

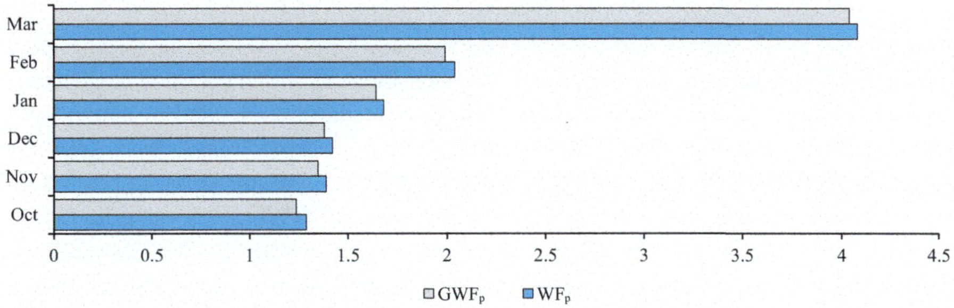

Fig. 18 Monthly average of GWF_P and WF_P (m^3/ton) (by authors)

consumption compared to the subsequent months, as illustrated in Fig. 18. Based on the results from November to December, the footprint indicators incrementally increased, when WF_P increased from 1.39 to 1.42 m^3/ton, and GWF_P from 1.35 to 1.38 m^3/ton. This trend continues during January, when WF_P is 1.68 m^3/ton and GWF_P is 1.64 m^3/ton, followed by the February's ratios of 2.04 m^3/ton for WF_P and 1.99 m^3/ton for GWF_P. The most notable increase occurs in March, where the WF_P ratio jumps to 4.08 m^3/ton and the GWF_P to 4.04 m^3/ton. This significant rise suggested a substantial escalation in water consumption and pollution, potentially due to increased production. Figure 18 shows that the percentage of GWF_P relative to WF_P peaked in March at 99%, indicating the highest pollution level. Conversely, October has the lowest ratio, and approximately 96% of WF_P is made up of GWF_P.

The efficiency of reducing GWF_P compared to WF_P has been insignificant (Fig. 18). In March, the minimum value was 1%, and in October, the maximum value was 4%. Moreover, the small proportion of BWF to the overall WF indicates that the production of products in the studied Industrial Park have polluted relatively more water than consuming freshwater.

4 Findings

This chapter initially introduced industrial GWF. Second, based on recent literature, it reviewed the advantages of GWF and pointed to the gaps and challenges of this index. For example, it is revealed that GWF has the privilege of evaluating the performance of WWTPs based on removing multiple pollutants. The GWF is complex because it covers different aspects of water quality management and requires a solid understanding of its fundamental concepts and accounting methods. One of the main challenges in industrial GWF accounting is rooted in the methodological uncertainties for assessing multiple pollutants, as well as the need to account for emerging pollutants that may necessitate modifications to the current methods. On the other hand, standards are very important for measuring GWF, as they guide the applied methods. These standards can be derived from global guidelines or tailored to specific regions. International standards typically make it easier to compare the final results, while local standards are relatively more accurate. Measuring GWF can be tricky as it depends on many factors, which can lead to different results and some uncertainties. This is why more research is needed to expand our knowledge about these uncertainties. It will help improve GWF as a tool for environmental monitoring, management, and decision-making. It is important to work on these areas to use GWF effectively. In addition, GWF has some potential to include the impaired status of the receiving environment in its calculations. According to this potential, GWF can be developed as a unique environmental index for regional or basin-scale water quality assessments. Accordingly, the possible impact assessment of WWTPs, for point emission sources, and best management practices (BMPs) for non-point emission sources, can be modeled and compared through GWF and its sustainability. However, GWF still has some gaps in both the concept and calculation method. For instance:

- An effluent with exceeding water quality standards due to the presence of hazardous contaminants and is transported for disposal requires separate GWF accounting at the actual discharge location.
- Industrial WWTPs work to ensure that the wastewater they release, known as effluent, has pollutant levels below certain environmental safety targets. However, the GWF, which measures the impact of pollutants on freshwater, can still be influenced by substances that the treatment plants do not fully remove or by new pollutants that are created during the treatment process. This means that even if the effluent meets the regulatory standards, there could still be environmental impacts that need to be considered and addressed.
- GWF index should reflect the environmental impact of water consumption, considering pollutant concentrations rather than just the volume used.
- GWF assessment relies on discharged volumes and concentrations. It makes GWF unique for each receiving stream and complicates comparing treatment units on larger scales.
- When assessing GWF, it is important to consider not just a single discharge, but also the cumulative effect of multiple discharges and runoff from irrigation. This helps to get a complete picture of the water quality and the potential environmental impact of industrial processes.
- C_{max}, the maximum pollutant concentration, is vital for calculating the GWF based on discharge locations and standards. It considers various environmental scenarios and pollutant effects. Accurate GWF assessment requires recalculating for different scenarios to reflect the true environmental impacts.
- The comprehensive evaluation of industrial GWF requires considering a broad range of pollutants to accurately estimate the type and amount of pollution.
- The determination of a negative GWF index (GWF < 0) is incorrectly considered in articles, while a value less than zero indicates that the water body does not have a diluting role but is diluted by the incoming wastewater.

- Using multiple water resources and having different discharge points complicates GWF calculations. Therefore, these conditions require separate assessments considering the impacts of different treatment units.
- Effluent reuse within the same unit can theoretically reduce GWF. However, this claim requires case-specific studies.
- Multi-pollutant GWF requires weighting methods which may shift the leading pollutants.
- GWF calculations must consider the actual flow rates and pollutant concentrations due to their variability rather than relying on historical or similar data.
- In GWF, it is essential to account for the quantity of both the water extracted and the wastewater produced. These two factors play a pivotal role in assessing the environmental impact.
- Additives used for industrial wastewater treatment can increase pollution indicators, potentially resulting in a higher GWF at the treatment unit's outlet than its inlet.
- C_{nat} is a key parameter for GWF. However, due to data scarcity, global constants are usually used, and it requires regional adjustments for more accuracy.
- The reliance of GWF on water quality examinations and various parameters introduces some uncertainties that make policy-making more challenging, despite global standards aiding for its uniformity.
- The ratio of the industrial GWF index in an industrial park per consumed water is required for performance evaluation and long-term decision-making at the national scale.

According to the aforementioned gaps and potentials, this chapter finally evaluated the industrial multi-pollutant GWF of an industrial park. Based on the experimental results of the study area to calculate the GWF, it is crucial to consider both the volume of abstracted water and the generated wastewater. The water withdrawal transferred to the Industrial Park increased from 2929 m^3/day in October to 3573 m^3/day in March. Additionally, the daily average volumetric flow rate of industrial effluent ranged from 1008 to 2040 m^3/day. Although the concentrations of pollutants have impacts on the GWF, the monthly GWF in March was mainly influenced by the flow rate. Specifically, based on the BOD parameter, the GWF was estimated at 5.074 MCM in March, while the average of the GWF during the studied period was computed at about 1.711 $\pm$ 0.691 MCM/month. The average GWF due to NH_4, as the leading pollutant, quantified at about 2.155 $\pm$ 0.285 MCM/month, represents a significant volumetric impact with potential ecological ramifications for the aquatic ecosystems.

Based on the monthly GWF, the maximum value for GWF_P was calculated. The highest GWF_P values were observed in March, reaching 4.04 m^3/ton for the Industrial Park. It represents the volume of fresh water required to virtually assimilate the discharged pollution from the studied Industrial Park for manufacturing 1 ton of product.

5 Future Trends and Innovations

There are three key perspectives in this study: (1) assessing and estimating the WF of industrial effluents, considering multiple sources; (2) the evaluation of GWF per unit of production; and finally (3) the evaluation of industrial GWF during various periods. In future, two approaches will be used to clarify the assessment of GWF: (1) determining the efficiency of GWF% based on both influent and effluent (e.g. by an industrial WWTP); (2) comparing the performance of treatment units for reducing the GWF with a focus on removal efficiency. These innovative approaches contribute to addressing the available challenges. GWF accounting can be approached through different methods, such as volumetric, stress-weighted, and LCA, each with its own set of challenges and ongoing updates to methodologies (Jamshidi 2021). Also, the challenges and uncertainties are associated with the assessment of environmental impacts, particularly in the context of life cycle impact assessment (LCIA) through recent databases and models like ReCiPe (2016). It can also be combined with basin simulation tools, like soil and water assessment tools (SWAT) for the GWF assessment of agricultural productions (Jamshidi and Naderi 2023). GWF is also capable of integrating with other environmental indices, like carbon footprint (which measures emitted carbon dioxide over time, known as CF), for more holistic evaluations. This allows policymakers to integrate water quality with energy consumption under the new insights on the water–energy nexus, water–food nexus, and the LCA of manufactured or farmed products.

Based on the results obtained from this study and the conducted analyses, the following recommendations can be proposed for the development of research related to the GWF index for industrial wastewater and contribute to sustainable water management practices:

1. Individual assessment for each industrial unit:
   - It is recommended to determine the industrial wastewater GWF for specific industrial enterprises. We also recommend assessing the impacts of WWTPs on each unit, separately for better comparison.
2. Comparison of treatment scenarios:
   - Reusing treated wastewater in industries can influence the GWF of the final output, and it is essential to evaluate these impacts to ensure sustainable water management practices.
3. Industrial wastewater reuse:
   - To calculate the GWF index more insightfully, consider using innovative methods like normalizing the index against annual production or energy use. This allows for a fair comparison of GWF among various industrial facilities, highlighting efficiency and sustainability practices.

4. Looking for uncertainties:
   - Expand the research to encompass other industrial complexes exhibiting similar uncertainties. To validate the findings, conduct preliminary field studies and experiments for comparative analysis.
5. Study GWF based on multiple pollutants:
   - GWF may differ across various units or treatment modules due to distinct parameters. However, calculating the final GWF value by simply summing up the footprints from different parameters is not accurate. A comprehensive assessment that considers the unique impact of each parameter is necessary for correct GWF evaluation.

Acknowledgments We sincerely appreciate Bina Azma Sepahan Laboratory for its expertise and diligent efforts in conducting a comprehensive analysis of industrial wastewater parameters for this study. We are also thankful for the financial support provided under contract number 1401/9475. Their commitment to accuracy and precision ensured the reliability of the data presented in this chapter. We also express our gratitude to the Iran Small Industries and Industrial Parks Organization (ISIPO) in Isfahan for their financial support throughout this research endeavor under Grant No. 37720/1/2430. Their collaboration and assistance were essential in facilitating access to the necessary resources and information. This study stands as a testament to the collective efforts of both organizations in advancing our understanding of wastewater management.

References

Ahmed J, Thakur A, Goyal A (2021) Industrial wastewater and its toxic effects. In: Shah MP (ed) Biological treatment of industrial wastewater. The Royal Society of Chemistry 1–14

Amuda OS, Amoo IA, Ajayi OO (2006) Performance optimization of coagulant/flocculant in the treatment of wastewater from a beverage industry. J Hazard Mater 129(1–3):69–72

Ansorge L, Stejskalová L, Dlabal J (2020a) Grey water footprint of point sources of pollution: Czech republic study. J Urban Environ Eng 14(1):144–149

Ansorge L, Stejskalová L, Dlabal J (2020b) Grey water footprint as a tool for implementing the Water Framework Directive – Temelín nuclear power station. J Clean Prod 263:121541

Ansorge L, Stejskalová L, Dlabal J (2020c) Effect of WWTP size on grey water footprint—Czech Republic case study. Environ Res Lett 15(10):104020

APHA (2012) Standard methods for the examination of water and wastewater, vol 10. American Public Health Association, Washington, DC

Appannagari RR (2017) Environmental pollution causes and consequences: a study. North Asian Int Res J Soc Sci Humanit 3(8):151–161

Arastou K (2024) Grey water footprint and quality evaluation of industrial wastewater by multiple pollutants (Case Study: Jey Industrial Park treatment plant), M.Sc thesis, Faculty of Civil Engineering and Transportation, University of Isfahan, Under the supervision of Dehnavi A and Jamshidi S

ASTM (2023) American Society for testing and materials. Available from: https://www.astm.org

Australian and New Zealand Council (2000) Australian and New Zealand guidelines for fresh and marine water quality, vol 3, Primary Industries — Rationale and Background Information

Baban A (2013) Biodegradability assessment and treatability of high strength complex industrial park wastewater. CLEAN–Soil, Air, Water 41(10):976–983

Babatsouli P, Palogos I, Michalodimitraki E, Costa C, Kalogerakis N (2015) Evaluation of a MBR pilot treating industrial wastewater with a high COD/N ratio. J Chem Technol Biotechnol 90 (1):26–33

Barakat MA (2011) New trends in removing heavy metals from industrial wastewater. Arab J Chem 4(4):361–377

Bay A, Sadeghi M, Rahimzade H, Behnampour N (2018) Survey of industrial wastewater treatment performance with biological monitoring of protozoa in the industrial park of abadan. Environ Eng Manage J 17(7):1679–1683

Billen G, Garnier J, Deligne C, Billen C (1999) Estimates of early-industrial inputs of nutrients to river systems: implication for coastal eutrophication. Sci Total Environ 243:43–52

Bisschops I, Spanjers H (2003) Literature review on textile wastewater characterisation. Environ Technol 24(11):1399–1411

Borsato E, Galindo A, Tarolli P, Sartori L, Marinello F (2018) Evaluation of the grey water footprint comparing the indirect effects of different agricultural practices. Sustainability 10(11):3992

Chang MR, Lee DJ, Lai JY (2007) Nanoparticles in wastewater from a science-based industrial park—Coagulation using polyaluminum chloride. J Environ Manage 85(4):1009–1014

Chapagain AK, Hoekstra AY (2008) The global component of freshwater demand and supply: an assessment of virtual water flows between nations as a result of trade in agricultural and industrial products. Water Int 33(1):19–32

Chapagain AK, Hoekstra AY, Savenije HHG, Gautam R (2006) The water footprint of cotton consumption: an assessment of the impact of worldwide consumption of cotton products on the water resources in the cotton producing countries. Ecol Econ 60(1):186–203

Chen Z, Sun Y, Ai X, Malik SM, Yang L (2019) Integrated demand response characteristics of industrial park: a review. J Mod Power Syst Clean Energy 8(1):15–26

Chislock MF, Doster E, Zitomer RA, Wilson AE (2013) Eutrophication: causes, consequences, and controls in aquatic ecosystems. Nat Educ Knowl 4(4):10

Cui S, Dong H, Wilson J (2020) Grey water footprint evaluation and driving force analysis of eight economic regions in China. Environ Sci Pollut Res 27:20380–20391

Department of Environment Islamic Republic of Iran (2016) Water quality standard in Iran

Department of Environment Islamic Republic of Iran (2023) Wastewater discharge standard in Iran

Deshpande DP, Patil PJ, Anekar SV (2012) Biomethanation of dairy waste. Res J Chem Sci 2231:606

Directorate-General for Health and Food Safety (European Commission) (2018) Technical guidance for deriving environmental quality standards. European Commission

Dong H, Zhang L, Geng Y, Li P, Yu C (2021) New insights from grey water footprint assessment: an industrial park level. J Clean Prod 285:124915

Environmental Protection Organization of Iran (1979) Measurement of water quality and biological study of water in the Zayandeh Rood river

Falkenmark M, Lindh G (1974) How can we cope with the water resources situation by the year 2015. Ambio 114–122

Fanchiang J, Tseng D, Guo G, Chen H (2009) Ozonation of complex industrial park wastewater: effects on the change of wastewater characteristics. J Chem Technol Biotechnol Int Res Process Environ Clean Technol 84(7):1007–1014

FAO (2023) Water quality evaluation. Available from: https://www.fao.org/3/t0234e/t0234e01.htm

Felicio M, Amaral D, Esposto K, Gabarrell Durany X (2016) Industrial symbiosis indicators to manage eco-industrial parks as dynamic systems. J Clean Prod 118:54–64

Fipps G (2003) Irrigation water quality standards and salinity management strategies. Texas FARMER Collection

Franke NA, Boyacioglu H, Hoekstra AY (2013) Grey water footprint accounting: Tier 1 supporting guidelines, vol 65. Unesco-Ihe Delft, Delft

Gaillardet J, Viers J, Dupré B (2003) Trace elements in river waters. Treatise on geochemistry 5:605

Gourgiotis A, Kyvelou SS, Lainas I (2021) Industrial location in Greece: fostering green transition and synergies between industrial and spatial planning policies. Land (Basel) 10(3):271

Gu Y, Dong Y, Wang H, Keller A, Xu J, Chiramba T et al (2016) Quantification of the water, energy and carbon footprints of wastewater treatment plants in China considering a water–energy nexus perspective. Ecol Indic 60:402–409

Guteta G, Worku H (2023) The role of industrial parks in managing sustainability challenges of urban transition: empirical analysis of the context in Adama and Hawassa Industrial Parks of Ethiopia. Environ Dev Sustain 1–27

Hanchang SHI (2009) Industrial wastewater-types, amounts and effects. Point sources of pollution: local effects and their control 2:191

Handayani W, Kristijanto AI, Hunga AIR (2019) A water footprint case study in Jarum village, Klaten, Indonesia: the production of natural-colored batik. Environ Dev Sustain 21(4):1919–1932

Hasehzadeh HR, Khakzand M, Ojaghlou M (2018) Optimal thermal characteristics of the courtyard in the hot and arid climate of Isfahan. Buildings 8(12):166

Health Ministry of New zealand Government (2019) Volume 3 Datasheets – chemical and physical determinands. Part 2.5: Aesthetic determinands

Henderson V, Kuncoro A, Turner M (1995) Industrial development in cities. J Polit Econ 103 (5):1067–1090

Herrebrugh RC (2018) The blue and grey water footprint of industry and domestic water supply: the blue and grey water footprint of all industrial sectors and domestic water supply for each country annually in the period 1960–2015

Hoekstra AY, Mekonnen MM (2012) The water footprint of humanity. Proceedings of the National Academy of Sciences 109(9):3232–3237. Available from: https://www.pnas.org/doi/abs/10.1073/pnas.1109936109

Hoekstra AY, Chapagain AK, Aldaya MM, Mekonnen MM (2011) The water footprint assessment manual: setting the global standard. Routledge

Hu Y, Huang Y, Tang J, Gao B, Yang M, Meng F et al (2018) Evaluating agricultural grey water footprint with modeled nitrogen emission data. Resour Conserv Recycl 138:64–73

Hu W, Tian J, Li X, Chen L (2020) Wastewater treatment system optimization with an industrial symbiosis model: a case study of a Chinese eco-industrial park. J Indus Ecol 24(6):1338–1351

Huang Y, Zhou B, Han R, Lu X, Li S, Li N (2020) China's industrial gray water footprint assessment and implications for investment in industrial wastewater treatment. Environ Sci Pollut Res 27:7188–7198

Institute of Standards and Industrial Research of Iran (2020) Drinking water – physical and chemical specifications ISIRI 1053

Jamshidi S (2021) Grey water footprint accounting, challenges, and problem-solving. In: Agroecological footprints management for sustainable food system. Springer, 247–271

Jamshidi S, Naderi A (2023) A quantitative approach on environment-food nexus: integrated modeling and indices for cumulative impact assessment of farm management practices. PeerJ 11:e14816

Jamshidi S, Imani S, Delavar M (2020) Impact assessment of best management practices (BMPs) on the water footprint of agricultural productions. Int J Environ Res 14:641–652

Jamshidi S, Imani S, Delavar M (2022) An approach to quantifying the grey water footprint of agricultural productions in basins with impaired environment. J Hydrol 606:127458

Jamshidi S, Farsimadan M, Mohammadi H (2024) A holistic approach for performance evaluation of wastewater treatment plants: integrating grey water footprint and life cycle impact assessment. Water Sci Technol 89(7):1741–1756

Johnson MB, Mehrvar M (2021) From field to bottle: water footprint estimation in the winery industry. Water footprint: assessment and case studies 103–136

Karandish F (2019) Applying grey water footprint assessment to achieve environmental sustainability within a nation under intensive agriculture: a high-resolution assessment for common agrochemicals and crops. Environ Earth Sci 78(6):200

Khan S, Malik A (2014) Environmental and health effects of textile industry wastewater. Environmental deterioration and human health: natural and anthropogenic determinants 55–71

Kumar PS, Prasanth SM, Harish S, Rishikesh M (2021) Industrial water footprint: case study on textile industries. Water footprint: assessment and case studies 35–60
Leflaive X, Witmer M, Martin-Hurtado R, Bakker M, Kram T, Bouwman L et al (2012) Water
Lei G, Ren H, Ding L, Wang F, Zhang X (2010) A full-scale biological treatment system application in the treated wastewater of pharmaceutical industrial park. Bioresour Technol 101(15):5852–5861
Liu J, Tang M (2018) Wastewater management approach in an industrial park. Water Sci Technol 2017(2):546–551
Lyu Y, Ye H, Zhao Z, Tian J, Chen L (2020) Exploring the cost of wastewater treatment in a chemical industrial park: model development and application. Resour Conserv Recycl 155:104663
Mekonnen M, Hoekstra AY (2011a) National water footprint accounts: the green, blue and grey water footprint of production and consumption. Volume 1: Main Report
Mekonnen M, Hoekstra AY (2011b) National water footprint accounts: the green, blue and grey water footprint of production and consumption. Volume 2: appendices
Mekonnen MM, Hoekstra AY (2015) Global gray water footprint and water pollution levels related to anthropogenic nitrogen loads to fresh water. Environ Sci Technol 49(21):12860–12868
Mekonnen MM, Hoekstra AY (2018) Global anthropogenic phosphorus loads to freshwater and associated grey water footprints and water pollution levels: a high-resolution global study. Water Resour Res 54(1):345–358
Meng X, Lu J, Wu J, Zhang Z, Chen L (2022) Quantification and evaluation of grey water footprint in Yantai. Water (Basel) 14(12):1893
Minister of Health Canada (2022) Guidelines for Canadian Recreational Water Quality
Ministry Environment And Water of Malaysia (2021) National water quality standards
Morera S, Corominas L, Poch M, Aldaya MM, Comas J (2016) Water footprint assessment in wastewater treatment plants. J Clean Prod 112:4741–4748
Mousazadeh M, Niaragh EK, Usman M, Khan SU, Sandoval MA, Al-Qodah Z et al (2021) A critical review of state-of-the-art electrocoagulation technique applied to COD-rich industrial wastewaters. Environ Sci Pollut Res 28(32):43143–43172
MSCI S (2020) Global industry classification standard (GICS) methodology. MSCI, New York
Munter R (2003) Industrial wastewater characteristics. The Baltic University Programme (BUP), Sweden
Muthu SS, Gardetti MA (2020) Sustainability in the textile and apparel industries: production process sustainability. Springer, 163-187
National Health and Medical Research Council (2008) Guidelines for Managing Risks in Recreational Water, Australian Government
OECD (2006) Data sheets for surface water quality standards, vol 1
OECD (2007) Proposed system of surface water quality standards for Moldova. París. Obtenido de. http://www.oecd.org/environment/outreach/38120922
Postel SL, Daily GC, Ehrlich PR (1996) Human appropriation of renewable fresh water. Science (1979) 271(5250):785–788
Roudbari MV, Dehnavi A, Jamshidi S, Yazdani M (2023) A multi-pollutant pilot study to evaluate the grey water footprint of irrigated paddy rice. Agric Water Manage 282:108291
Rumelt RP (1991) How much does industry matter. Strateg Manag J 12(3):167–185
Sirianuntapiboon S, Jeeyachok N, Larplai R (2005) Sequencing batch reactor biofilm system for treatment of milk industry wastewater. J Environ Manage 76(2):177–183
Slavov AK (2017) General characteristics and treatment possibilities of dairy wastewater–a review. Food Technol Biotechnol 55(1):14–28
Sosnovskikh S (2017) Industrial clusters in Russia: the development of special economic zones and industrial parks. Russ J Ecol 3(2):174–199
Stejskalová L, Ansorge L, Kučera J, Vološinová D (2021) Grey water footprint as a tool for wastewater treatment plant assessment–Hostivice case study. Urban Water J 18(10):796–805

Swistock B (2016) Interpreting irrigation water tests. Basic interpretation of how various water quality parameters can influence plant growth during irrigation. Pennsylvania State University
Taiwan EPA (2022) Taiwan water quality standards, available data & initial analysis
Tedd K, Coxon C, Misstear B, Daly D, Craig M, Mannix A et al (2017) Assessing and developing natural background levels for chemical parameters in Irish groundwater. Environmental Protection Agency, Wexford, 183
Tian J, Liu W, Lai B, Li X, Chen L (2014) Study of the performance of eco-industrial park development in China. J Clean J Clean Prod 64:486–494
UN Environment Programme GEMStat (1980a) Zayandeh River in Isfahan. Available: 2024-04-17. Available from: https://gemstat.bafg.de/applications/public.html?publicuser=PublicUser#gemstat/Stations
UN Environment Programme GEMStat (1980b) Well No.4 in Isfahan. Available: 2024-04-17. Available from: https://gemstat.bafg.de/applications/public.html?publicuser=PublicUser#gemstat/Stations
Vice President's Planning and Strategic Supervision (2010) Environmental criteria for reuse of returned water and wastewaters in Iran, Publication No. 535
von Sperling M, de Lemos Chernicharo CA (2005) Biological wastewater treatment in warm climate regions. International Water Association Publications, 1, 810
Wang L, Ding X, Wu X (2013) Blue and grey water footprint of textile industry in China. Water Sci Technol 68(11):2485–2491
Wang S, Lu C, Gao Y, Wang K, Zhang R (2019) Life cycle assessment of reduction of environmental impacts via industrial symbiosis in an energy-intensive industrial park in China. J Clean Prod 241:118358
Weerasooriya RR, Liyanage LPK, Rathnappriya RHK, Bandara WBMAC, Perera TANT, Gunarathna MHJP et al (2021) Industrial water conservation by water footprint and sustainable development goals: a review. Environ Dev Sustain 23(9):12661–12709
Xueyi C, Meyer AN (2011) Management models of industrial parks in China. In: Overview of the current situation and draft concept for an international forum
Yan F, Kang Q, Wang S, Wu S, Qian B (2021) Improved grey water footprint model of noncarcinogenic heavy metals in mine wastewater. J Clean Prod 284:125340
Yapıcıoğlu PS (2019a) Grey water footprint of a dairy industry wastewater treatment plant: a comparative study. Water Pract Technol 14(1):137–144
Yapıcıoğlu P (2019b) Seasonal water footprint assessment for a paint industry wastewater treatment plant. Sakarya Univ J Sci 23(2):175–183
Yapıcıoğlu P, Yeşilnacar Mİ (2022) Grey water footprint assessment of groundwater resources in southeastern Turkey: effect of recharge. Water Supply 22(1):615–627
Yaseen DA, Scholz M (2019) Textile dye wastewater characteristics and constituents of synthetic effluents: a critical review. Int J Environ Sci Technol 16:1193–1226
Yazdani M (2019) The economic and statistical evaluation of climatic elements in Tabriz and Isfahan (in Iran). J Bus Manage Econ Res 3:18
Zhang L, Dong H, Geng Y, Francisco MJ (2019) China's provincial grey water footprint characteristic and driving forces. Sci Total Environ 677:427–435
Zhu R, Yang C, Zhou M, Wang J (2015) Industrial park wastewater deeply treated and reused by a novel electrochemical oxidation reactor. Chem Eng J 260:427–433

Assessing the Sustainability of Agricultural Water Use Based on Water Footprints of Wheat and Rice Production

Shahla Dehghanpir, Ommolbanin Bazrafshan, Sina Nadi, and Sajad Jamshidi

1 Introduction

The optimal and sustainable utilization of water resources is fundamental to the continuity of social and economic development, as well as to meeting the requirements of environmental management. The agriculture sector is the predominant consumer of water, accounting for 90% of global water consumption, a demand that has been increasingly on the rise worldwide (Hoekstra et al. 2011). Despite water being a renewable resource, its availability fluctuates both temporally and spatially. Consequently, there is an expanding discrepancy between the escalating demand and the finite nature of water resources, highlighting a critical challenge for sustainability (Elliott et al. 2014; Abdollahzadeh Kahrizi et al. 2022).

The escalation in water demand, particularly within the agricultural sector, has intensified the competition for freshwater resources. This has heightened concerns about water scarcity, increasingly attributed to agricultural activities (Gerbens-Leenes et al. 2012). Consequently, enhancing water productivity in agriculture and mitigating water stress caused by farming practices are crucial steps towards the sustainable management of water resources in any given region (Hoekstra and Mekonnen 2012). The examination of the interplay between agricultural output

S. Dehghanpir · O. Bazrafshan (✉)
Department of Natural Resources, University of Hormozgan, Bandar Abass, Iran
e-mail: o.bazrafshan@hormozgan.ac.ir

S. Nadi
Department of Mathematics, Purdue University, West Lafayette, IN, USA
e-mail: snadi@purdue.edu

S. Jamshidi
Department of Agronomy, Purdue University, Lafayette, IN, USA
e-mail: sjamshi@purdue.edu

S. S. Muthu (ed.), *Sustainability and Water Footprint*, Environmental Footprints and Eco-design of Products and Processes, https://doi.org/10.1007/978-3-031-70810-7_3

and water resources has long been a global issue of interest, especially pronounced in arid and semi-arid regions, underscoring the urgent need for sustainable water management strategies (Wang et al. 2015; Cao et al. 2014).

In light of these considerations, contemporary efforts to explore the relationship between agricultural production and available water resources (AWRs) employ a variety of indicators. These include the water footprint (WF) concept, water productivity, water critical ratio, water stress index (WSI), and water poverty. Among these, the concept of WF stands out as a particularly crucial indicator designed to monitor the scarcity of water resources. This approach enables a comprehensive assessment of water use (WU) and management in agriculture, offering insights into the efficiency and sustainability of water utilization in the production process.

The concept of WF was introduced in 2002 by Hoekstra and Hung (2002) and has since been applied extensively in the management of freshwater resources. The WF of a product represents the total volume of freshwater utilized in its production, encompassing the entirety of the product's production chain. This concept is broken down into three components: blue, green, and gray WFs (Fu et al. 2019). The blue WF pertains to the consumption of water resources from surface and groundwater sources during the production process. The green WF is associated with the use of green water resources, and it primarily refers to rainwater. Meanwhile, the gray WF measures the volume of freshwater required to assimilate pollutants, reflecting the environmental impact of production processes.

Besides WF, WSI is another pivotal metric widely utilized in examining water scarcity. Initially proposed by Falkenmark in 1989, this index is calculated as the ratio of the amount of water used for product production to the AWR. Falkenmark's classification delineates five levels of water stress that a region can experience: no water stress, low water stress, medium water stress, high water stress, and very high water stress (Falkenmark 2001). This framework allows for a nuanced understanding of the intensity of water scarcity in different regions, facilitating targeted and effective water management strategies.

The concept of water poverty, another critical indicator for assessing the status of water resources in a region, was first introduced by Sullivan in 2002 in the context of water scarcity. This index is determined by the interaction between the total WF in the agricultural sector and the water resources available in a region (Sullivan 2002). Numerous researchers have employed these two indicators—water poverty and WSI—alongside considerations of the WF and regional water availability to evaluate water shortages in agricultural production. Notable studies in this area include Cao et al. (2017), Cao et al. (2018), and Liu et al. (2017), among others. These investigations underscore the utility of integrating multiple indicators to provide a comprehensive assessment of water scarcity challenges faced by agricultural sectors worldwide.

Iran, characterized by its arid and semi-arid climate, is grappling with a critical water shortage that has profound and, in some instances, irreversible effects on its economy, ecosystem functions, and the well-being of its populace (Jamshidi et al. 2019a, b). The country's unique topographical and climatic conditions result in an unfavorable temporal and spatial distribution of rainfall, posing significant challenges to sustainable crop production. The agricultural sector emerges as a primary and voracious consumer of water resources in Iran, accounting for over 90% of the nation's freshwater usage. This underscores the importance of understanding the allocation of water resources for agricultural production for managers and policymakers. Addressing water scarcity, therefore, necessitates a strategic reduction in water consumption within the agricultural sector (Madani 2014). Previous studies have shown the importance of WF assessments in arid and semi-arid regions such as in Iran (Rafiei Sardooi et al. 2024; Fathian et al. 2023; Bazarfshan et al. 2022).

Wheat and rice stand out as globally significant agricultural products, with wheat's extensive consumption and demand marking it as a strategic commodity. In Iran, wheat is pivotal for food security, influencing the cultivation patterns of other crops (Hadi et al. 2017). Following wheat, rice is recognized as a crucial staple and a strategic product for developing countries, including Iran, playing a vital role in food security (Ebrahimi Mahmoudi et al. 2021; Jamshidi et al. 2023). Hence, this study aims to apply the concept of WF to evaluate the water stress and water poverty induced by the production of wheat and rice. These measures serve as analytical tools for assessing water shortages in the producing provinces of Iran, highlighting the need for strategic water resource management in the agricultural sector to safeguard food security and ecosystem sustainability.

In this research, at first, we introduced the methodology of sustainability indicators of WU in the agricultural sector based on the WF and then introduced the data used. We divided the results into two parts; the first part presents all the indicators for the wheat production, and the second part presents the sustainability indicators of WU of rice production. Finally, we summarize the results.

2 Methods

WSI is defined as the ratio of WU to AWR, measured in cubic meters (Raskin et al. 1997; Hoekstra et al. 2011).

$$\mathrm{WSI} = \frac{\mathrm{WU}}{\mathrm{AWR}} \tag{1}$$

The agricultural WSI (AWSI) within a region is determined by the ratio of the total agricultural WF (AWF) to the availability of agricultural water resources (AgWR) over a specific period. Consequently, the AWSI for the production of agricultural products, which utilizes both blue and green water resources in accordance with the WF framework, is calculated as follows.

$$\mathrm{AWSI} = \frac{\mathrm{AWF}}{\mathrm{AgWR}} \tag{2}$$

where AWF represents the WF of agricultural products, encompassing the aggregate WFs of all product types, measured in cubic meters. AWR refers to the water resources allocated for the production of agricultural products within a region.

The wheat (rice) water stress index (WWSI (RWSI)) within a region is established by the ratio of the total AWF for wheat (rice) to the available water resources (AWR) during a specified period. Thus, the WWSI (RWSI) for the production system of the wheat (rice) crop, grounded in the framework of blue and green water resources and WF, is determined as follows: AWF is the WF of wheat (rice) production, and AgWR denotes the water resources available for agriculture in the region.

$$\text{WWSI or RWSI} = \frac{\text{AWF}}{\text{AWR}}. \tag{3}$$

The term AWR encompasses the availability of both blue and green water resources. This is calculated by summing the availability of blue water resources (AWR_{Blue}) and the availability of green water resources ($\text{AWR}_{\text{Green}}$), as follows:

$$\text{AWR} = \text{AWR}_{\text{blue}} + \text{AWR}_{\text{green}} \tag{4}$$

$$\text{AWR}_{\text{blue}} = \text{WR} \times \frac{\text{AWU}}{\text{WU}} \tag{5}$$

WR denotes the total availability of water resources within a region, measured in cubic meters. The term agricultural water use (AWU) refers to the volume of water resources consumed within the agricultural sector, while WU signifies the overall consumption of water resources across all sectors in the region. $\text{AWR}_{\text{green}}$, representing the availability of green water resources, is calculated based on the effective annual rainfall received by agricultural lands. The specific formula used for calculating $\text{AWR}_{\text{green}}$ involves quantifying this effective rainfall, taking into account factors such as evapotranspiration, infiltration, and water retention capacity of the soil, to accurately assess the volume of rainwater beneficially used by crops.

$$\text{AWR}_{\text{green}} = 10 \times A \times P_e \tag{6}$$

where A is the area of arable land (m^2) and P_e is the annual effective rainfall (mm). The estimation of P_e is conducted following the methodology recommended by the FAO CROPWAT model.

$$P_e = \begin{Bmatrix} P\left(\frac{4.17 - 0.02 \times P}{4.17}\right), & P < 83 \\ (41.7 + 0.1 \times P), & P \geq 83 \end{Bmatrix} \tag{7}$$

Table 1 describes the classification of WSIs, agricultural water stress, WWSI, and RWSI (Pfister et al. 2009; Cao et al. 2017; Dehghanpir et al. 2024).

Table 1 Classification of water stress indicators, agricultural water stress, wheat and rice water stress indicators

Category	WSI	AWSI	WWSI, RWSI
No water stress	0–0.15	0–0.15	0–0.15
Low stress	0.16–0.3	0.16–0.3	0.16–0.3
Medium stress	0.31–0.6	0.31–0.6	0.31–0.6
High stress	0.61–0.80	0.61–1.2	0.61–1.2
Very high stress	0.81–0.9	1.2–1.9	1.2–1.9
Extreme stress	0.91–1	>1.9	>1.9

The AWF corresponds to the aggregate WF of all crop types, measured in cubic meters. This calculation consolidates the water consumed in the cultivation of various crops, providing a comprehensive metric of water usage within the agricultural sector:

$$\mathrm{AWF} = \sum_{i=1}^{n} (\mathrm{CWF}_i \times G_i) \tag{8}$$

where i is the number of crops, CWF shows the total WF for all types of crops, and G_i shows the amount of crop production in tons.

Regional CWF is calculated as the sum of WFs for all product types using the following equation:

$$\mathrm{CWF} = \mathrm{WF}_{\mathrm{blue}} + \mathrm{WF}_{\mathrm{green}} + \mathrm{WF}_{\mathrm{grey}} \tag{9}$$

Hoekstra et al. (2011) established the standard methodology for calculating the blue, green, and gray WFs within a crop production system. The method for calculating the gray WF is as follows:

$$\mathrm{WF}_{\mathrm{Grey}} = \frac{a \times \mathrm{NAR}}{C_{\mathrm{MAX}} - C_{\mathrm{NAT}}} \times \frac{1}{Y} \tag{10}$$

where a is a fraction of leaching, NAR is the amount of chemical fertilizer used in the field in kg/ha, C_{MAX} is the maximum acceptable concentration for nitrogen in kg/m^3, and C_{NAT} is the actual concentration of nitrogen in kg/m^3, which is considered as zero due to the non-existence of this value in Iran.

The blue WF encompasses the volume of surface and groundwater utilized throughout the production process. P_{e} constitutes the aggregate of effective rainfall received during the plant growth stage, measured in millimeters. $\mathrm{ET_C}$ refers to the combined plant evaporation and transpiration, also in millimeters, while Y signifies the yield of the crop, quantified in tons per hectare.

$$\mathrm{WF_{blue}} = \frac{(\mathrm{ET_c} - P_\mathrm{e}) \times 10}{Y} \tag{11}$$

Crop evapotranspiration ($\mathrm{ET_C}$) is calculated (mm/day) throughout the crop growth period. It is derived from the multiplication of the potential evapotranspiration of the reference plant ($\mathrm{ET_0}$), measured in millimeters per day, by the plant coefficient (K_C).

$$\mathrm{ET_c} = K_\mathrm{c} \times \mathrm{ET_0} \tag{12}$$

In this study, reference evapotranspiration was determined through the FAO-Penman Monteith method, utilizing the CROPWAT software, as outlined by Allan (1998).

$$\mathrm{ET_0} = \frac{0.408\Delta(R_\mathrm{n} - G) + \gamma \times \frac{900}{T+273} \times u_2 \times (e_s - e_a)}{\Delta + \gamma(1 + 0.34u_2)} \tag{13}$$

where $\mathrm{ET_0}$ is the standardized evapotranspiration for clipped grass or reference crop (mm day^{-1}); R_n represents the net radiation on the crop surface (MJ M^{-2} day^{-1}); G is the soil heat flux density (MJ M^{-2} day^{-1}) at the soil surface; T indicates the average temperature in degrees of Celsius; u_2 is the wind speed measured at a height of 2 m (ms^{-1}); e_s denotes the saturated vapor pressure ($\mathrm{kP_a}$); e_a is the actual vapor pressure ($\mathrm{kP_a}$); Δ represents the slope of the saturation vapor pressure–temperature curve ($\mathrm{KP_a}$ $^\circ\mathrm{C}^{-1}$); and γ is the psychrometric constant ($\mathrm{kP_a}$ $^\circ\mathrm{C}^{-1}$).

The green WF ($\mathrm{WF_{Green}}$) refers to the volume of rainwater consumed, which is calculated using the following equation.

$$\mathrm{WF_{Green}} = \frac{P_\mathrm{e} \times 10}{Y} \tag{14}$$

Water Deprivation Agriculture Sector

Water poverty is calculated by multiplying WF with WSI, and the result is expressed in cubic meters. This index gauges the impact of water utilization for crops such as wheat and rice relative to other agricultural products. A lower value of this index indicates lesser impact on the overall water consumption.

$$\text{Water Deprivation} = \mathrm{WF} \times \mathrm{WSI} \tag{15}$$

Water Dependence Index

Water dependence (WD) index is the ratio of the total net virtual water import (NVWI) to the total water allocated for the production of agricultural products (WU). This index shows the degree of dependence on foreign water sources through virtual water imports, which is calculated using Eq. (15). The range of this index is between zero and 100. A value of zero indicates a balance in the export and gross import of virtual water, and a value of 100 indicates complete dependence on the import of virtual water.

$$WD = \frac{NVWI}{WU + NVWI} \times 100 \tag{16}$$

Water Self-Sufficiency Index

The water self-sufficiency (WSS) index contrasts with the water dependence index by illustrating a region's capability to supply the water required for its domestic production. Should this index reach a value of 100, it signifies that the region possesses sufficient water resources to meet the demands of producing goods and services within its own boundaries, thereby eliminating the need for importing virtual water resources. This index is a measure of a nation's ability to furnish the water necessary for domestic production. As the index nears zero, it indicates substantial reliance on the importation of virtual water. A flowchart illustrating this methodology is depicted in Fig. 1.

$$WSS = 100 - WD \tag{17}$$

3 Data Sources

In this study, a comprehensive dataset was meticulously assembled to explore the intricate dynamics between agricultural practices and water resources in Iran, focusing particularly on wheat and rice production. The foundation of our research was laid upon detailed agricultural statistics, which encompassed the expanse of land under cultivation, the volume of crops produced, productivity per hectare, the timing of agricultural activities, and the intensity of fertilizer usage, all provided by the Ministry of Agriculture. Parallel to the agricultural data, an extensive collection of climatic records from the Iran Meteorological Organization offered insights into the environmental backdrop of crop production. This dataset included measurements of rainfall, humidity levels, solar exposure, and the fluctuation of temperatures, alongside wind speed observations across a network of meteorological stations, thereby painting a vivid picture of the climatic conditions that shape agricultural outputs. Demographic patterns and consumption behaviors within Iranian provinces also formed a critical component of our analysis. The Iranian Statistics Center supplied data on population distributions and the per capita intake of wheat and rice, shedding light on the demand side of the agricultural equation. Lastly, an examination of water resources, derived from the Iranian Water Resources Management, detailed the availability and utilization of both surface and subterranean water supplies. This analysis not only quantified total water consumption but also delineated the specific volumes dedicated to agriculture, thus highlighting the sector's pivotal role in

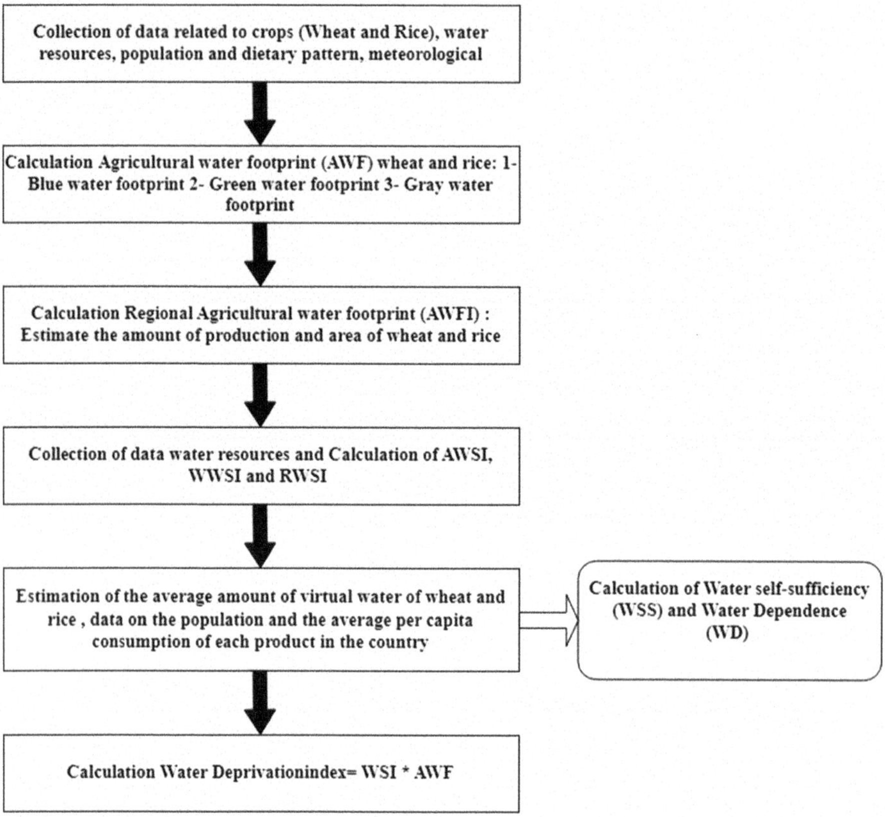

Fig. 1 Summary of the steps taken in current study to evaluate water footprints of wheat and rice production in Iran. (Ref: Authors)

national water dynamics. Illustrated in Fig. 2, a visual representation captures the fluctuation of water resources throughout the period from 2008 to 2019, offering a temporal perspective on the challenges and trends in water management in Iran's agricultural sector.

4 Results for Wheat

4.1 Statistics on Irrigated and Rain-Fed Wheat Cultivation

Wheat, recognized as a resilient and significant crop within the agricultural sector, is cultivated across all 32 provinces of Iran. The diverse climatic conditions across the country have positioned wheat as the predominant agricultural product in most provinces. In Iran, wheat cultivation is practiced under both irrigated and rain-fed conditions, reflecting its pivotal role in the nation's agriculture. The total area

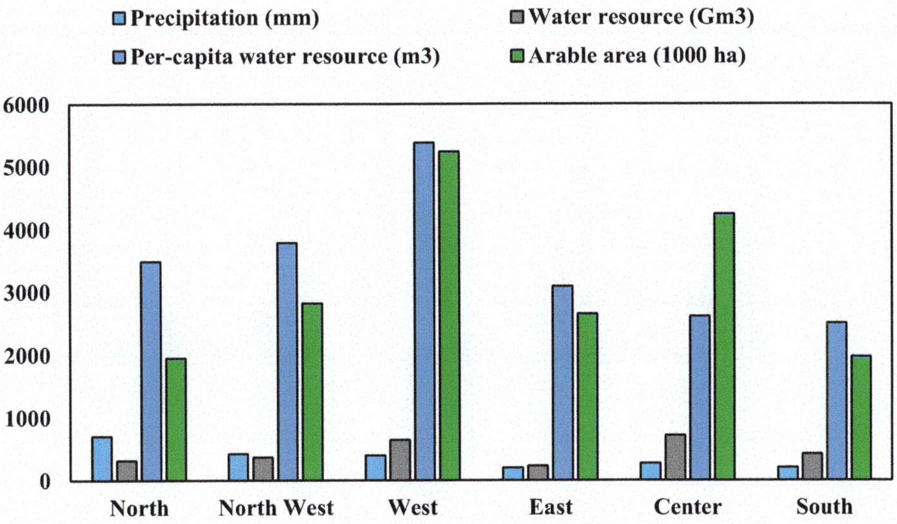

Fig. 2 The average of consumed water resources in different regions of Iran during the period of 2008–2019. (Dehghanpir et al. 2023)

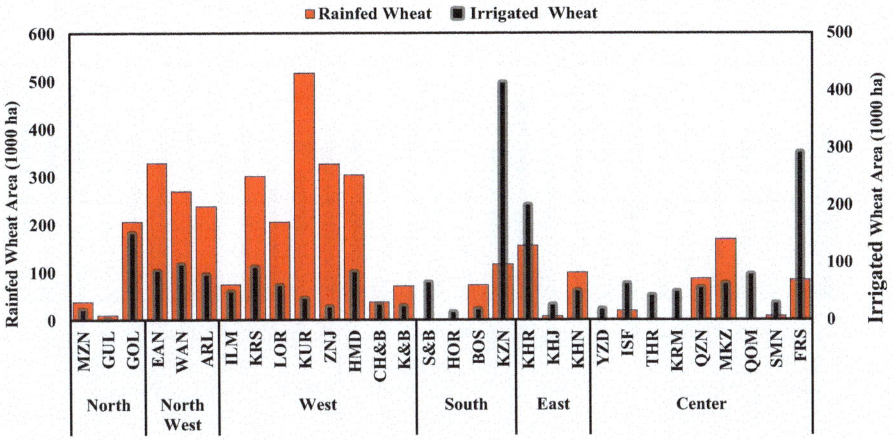

Fig. 3 The average cultivated area of irrigated and rain-fed wheat in Iran from 2008 to 2019. (MAJ 2021)

dedicated to wheat farming spans approximately 6.2 million hectares, with irrigated wheat fields covering about 2.2 million hectares, or 35%, and rain-fed wheat areas extending over 4 million hectares, accounting for 65%. Data detailing the average cultivated area, production volumes, and yields for wheat under both irrigated and rain-fed conditions from 2008 to 2019 are illustrated in Figs. 3, 4 and 5.

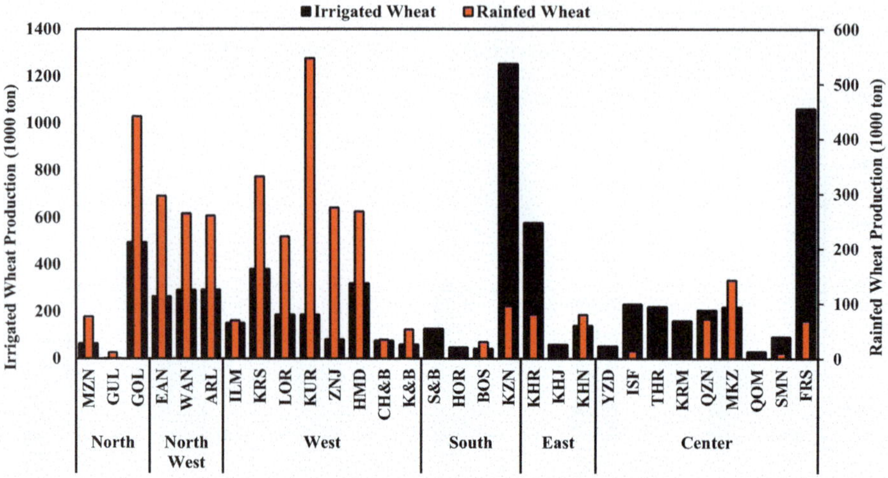

Fig. 4 The average production of irrigated and rain-fed wheat in Iran from 2008 to 2019. (MAJ 2021)

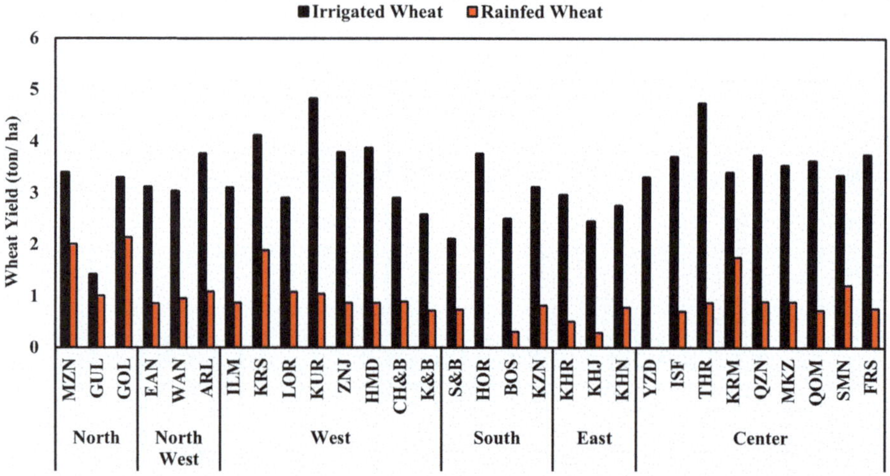

Fig. 5 The average crop yield of irrigated and rain-fed wheat in Iran from 2008 to 2019. (MAJ 2021)

4.2 *Water Footprint of Irrigated and Rain-Fed Wheat Production in Iran*

The analysis of WF components in wheat production across Iran, as depicted in Fig. 6 for the period of 2008 to 2019, reveals that the blue WF constitutes the predominant portion of WFs, succeeded by green and gray WFs. The study highlights that, within the realm of irrigated wheat cultivation, Khuzestan, Fars, and

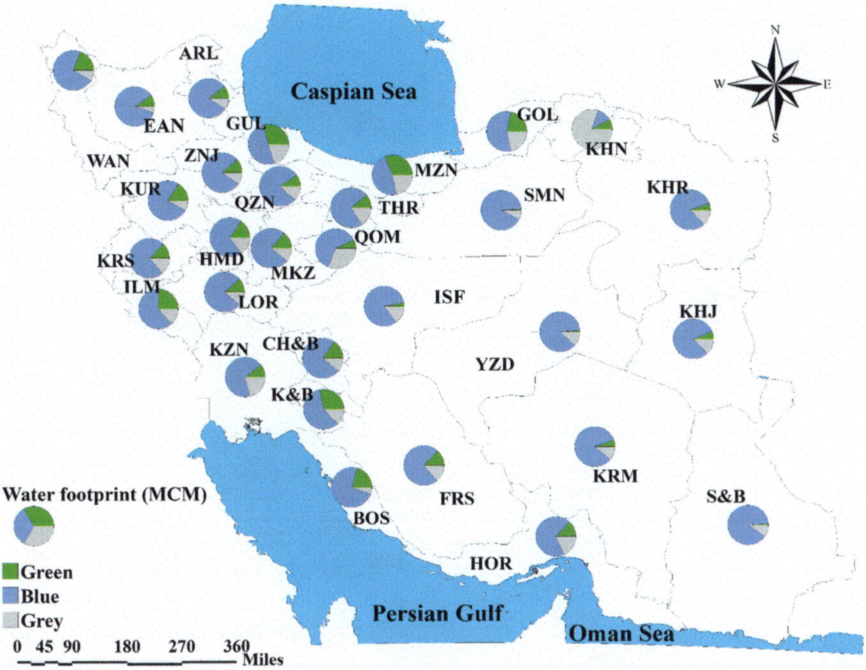

Fig. 6 The average volume of water footprint components of irrigated wheat in Iran from 2008 to 2019. (Ref: authors)

Razavi Khorasan provinces exhibit the highest blue WFs, registering at 1922.4, 1875.67, and 1464.92 million cubic meters (MCM), respectively. These figures correlate with the most significant agricultural outputs in the southern, central, and eastern regions of Iran. Conversely, Guilan province, located in the north of Iran, records minimal blue WF at 0.21 MCM, aligning with its lower production levels.

Additionally, the data indicates that Khuzestan and Gilan provinces have the highest and lowest gray WFs, at 606.97 and 9.09 MCM, respectively. Fars province, situated in central Iran, has the highest green WF at 336.17 MCM, while Guilan province, in the north, has the lowest at 0.12 MCM.

In the context of irrigated wheat, the blue WF emerges as the most substantial, averaging 73.50%. This suggests that, in irrigated wheat cultivation, blue water resources (comprising surface and underground water) are utilized more extensively than green water sources (effective rainfall). The contributions of green and gray WFs to the total WF of irrigated wheat stand at 13% and 14%, respectively.

The changes of WF components (green and gray WFs) in rain-fed wheat from 2008 to 2019 in each of the studied provinces are shown in Fig. 7. According to the figure, in the northern and western regions of Iran, the green WF is greater than the gray WF. While in the southern, central, and eastern provinces, on average, the gray WF is greater than the green WF. In Iran, for rain-fed wheat, the share of green and gray WF is 55% and 45%, respectively. Among the studied provinces, Kurdistan has

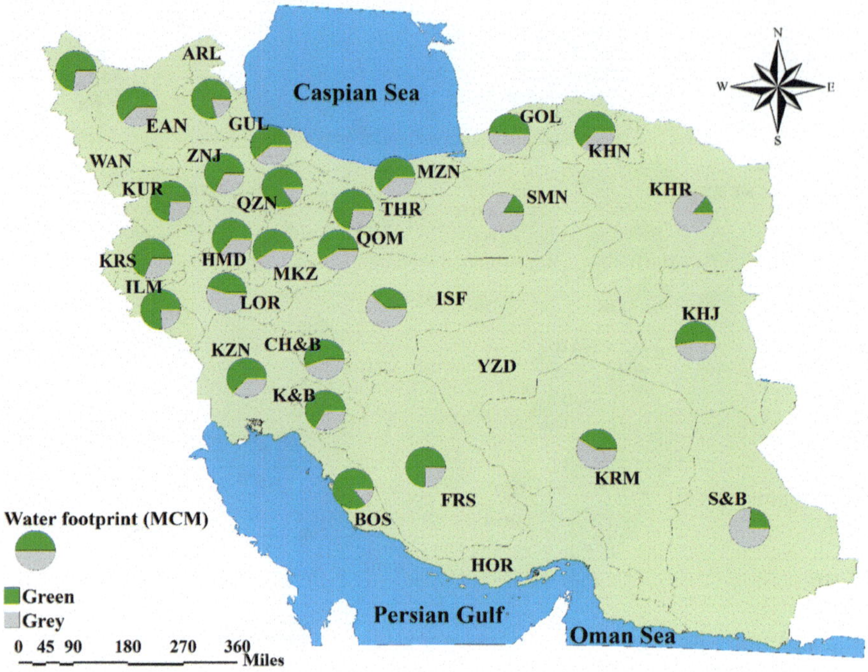

Fig. 7 The average volume of water footprint components of rain-fed wheat in Iran from 2008 to 2019. (Dehghanpir et al. 2022)

the largest green WF, with a value of 668.50 MCM. Hormozgan and Yazd provinces have no share due to the lack of rain-fed wheat. Kurdistan has the largest footprint of gray water with 239.76 MCM.

4.3 Water Stress Index in Agricultural Sector (AWSI) and Water Stress in Wheat Production (WWSI) in Iran

The analysis of WF components, specifically green and gray WFs, for rain-fed wheat across various provinces in Iran from 2008 to 2019 is illustrated in Fig. 8. The findings indicate a regional variation in the predominance of WF components. In the northern and western provinces of Iran, the green WF, which accounts for water from precipitation that is used by crops, exceeds the gray WF, associated with the volume of freshwater needed to assimilate the pollutants. Conversely, in the southern, central, and eastern provinces, the gray WF, on average, surpasses the green WF.

Nationally, for rain-fed wheat, the distribution between green and gray WFs stands at 55% and 45%, respectively. Among the provinces analyzed, Kurdistan emerges with the highest green WF, amounting to 668.50 MCM. Notably,

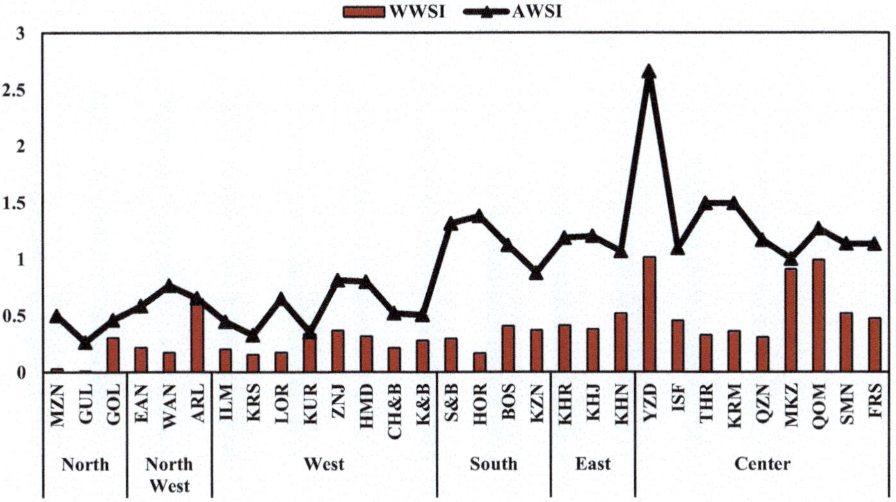

Fig. 8 Comparison of AWSI and WWSI during the period from 2008 to 2019 at the provincial scale. (Ref: Authors)

Hormozgan and Yazd provinces do not contribute to the green WF for rain-fed wheat due to the absence of such cultivation in these regions. Additionally, Kurdistan is identified as having the most significant gray WF, totaling 239.76 MCM. This differentiation underscores the varying reliance on green and gray water resources across Iran's agricultural landscape, influenced by regional climatic and agricultural practices.

4.4 *Temporal Variation of AWSI and WWSI in Wheat Cultivation*

The temporal variation of the AWSI and the WWSI, as depicted in Fig. 9, reveals critical insights into water stress levels over the observed period. The average AWSI during this timeframe stood at 0.92, consistently exceeding 0.8. This indicates that the production of agricultural products in Iran has been subject to severe water stress. Moreover, the AWSI has remained relatively stable without significant fluctuations.

On the other hand, the WWSI was recorded at 0.6, positioning it within the medium-to-high stress category according to the classification provided in Fig. 9. Similar to AWSI, the WWSI did not exhibit a notable trend over time, suggesting a persistent level of water stress affecting wheat production specifically. These findings underscore the challenges faced in managing water resources for agriculture in Iran, highlighting the need for effective strategies to mitigate water stress in both general agricultural and specific crop production contexts.

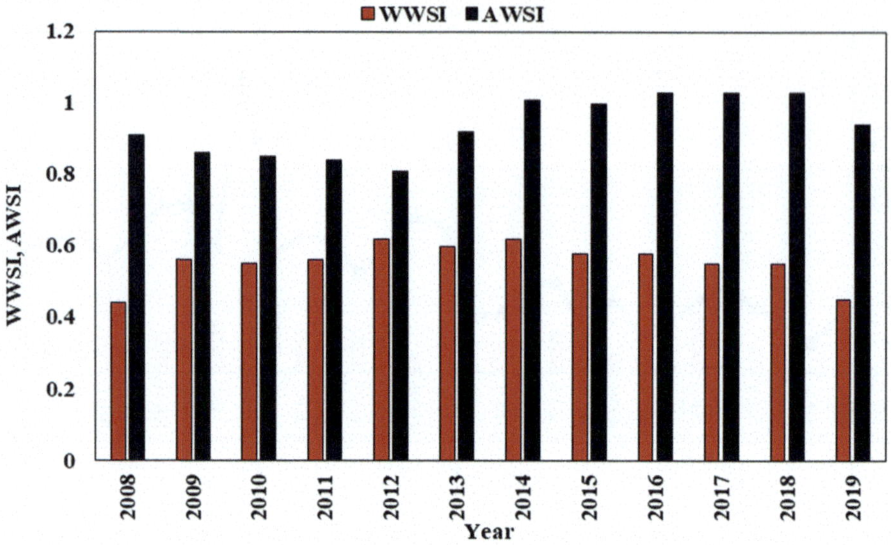

Fig. 9 Temporal variation of AWSI and WWSI in Iran during the period 2008–2019. (Ref: Authors)

4.5 *Water Deprivation in Wheat Cultivation at the National and Provincial Scales*

Water deprivation, defined as the product of the WF and the Agricultural AWSI, varies across regions, with higher WFs leading to greater levels of water deprivation. The temporal changes in wheat water deprivation from 2008 to 2019 for each province, as shown in Fig. 10, reveal significant disparities across Iran. Guilan province, with a water deprivation of 0.34 MCM, experiences the least water poverty, while Ardabil province faces the most severe water poverty, with a value of 2801.49 MCM. Following Ardabil, Kurdistan, Golestan, Kermanshah, and Fars provinces exhibit high levels of water deprivation, with values of 2006.17, 1349.78, 1151.45, and 1004.44 MCM, respectively.

This distribution indicates that the northeastern and western provinces of Iran, due to their extreme water stress levels and substantial WFs in wheat production, endure the highest amounts of water deprivation. Conversely, regions with smaller agricultural areas and lower water consumption for wheat cultivation exhibit lesser degrees of water deprivation. This pattern underscores the critical impact of water stress and agricultural practices on regional water scarcity, highlighting the importance of sustainable water management in mitigating water poverty in the wheat-producing provinces.

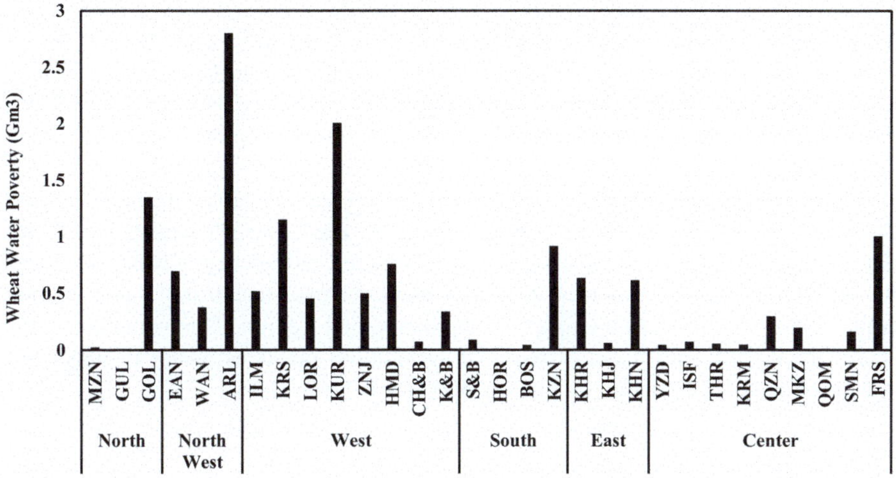

Fig. 10 Temporal variation of wheat water deprivation at the provincial scale during the period 2008–2019. (Ref: Authors)

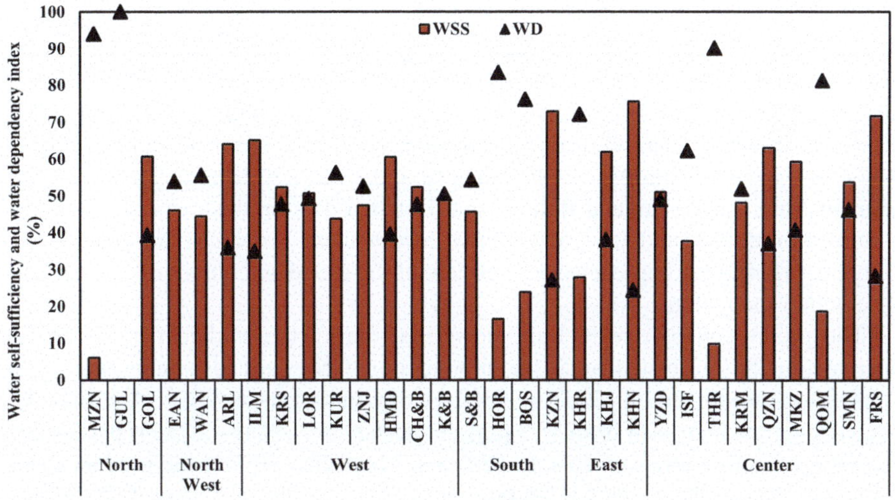

Fig. 11 Wheat water self-sufficiency and water dependency at the provincial scale in Iran. (Ref: Authors)

4.6 Wheat Water Self-Sufficiency and Water Dependency

Figure 11 illustrates the water dependency (WD) index and WSS index across various provinces, revealing a direct correlation between wheat production volumes, population size, and WSS levels. Provinces with higher wheat output and smaller consumer populations exhibit elevated self-sufficiency indices. Notably, North

Khorasan, Khuzestan, and Fars provinces boast the highest levels of WSS (and correspondingly, the lowest dependency indices), indicating a greater degree of autonomy in water resource utilization for wheat production. Conversely, Guilan, Mazandaran, and Tehran provinces display the lowest self-sufficiency indices, reflecting the highest dependency on virtual water imports for agricultural production.

These findings highlight the stark contrast in water resource management and dependency on external water sources among the provinces. Provinces like North Khorasan, Khuzestan, and Fars achieve over 70% self-sufficiency in water resources for wheat cultivation, underscoring their lesser reliance on virtual water imports compared to provinces like Guilan, Mazandaran, and Tehran. This variance underscores the significant impact of regional agricultural practices, water management strategies, and demographic factors on WD and WSS in Iran.

5 Results for Rice

5.1 Average Cultivated Area, Production, and Yield of Rice in Iran

Rice stands as a prized and extensively consumed staple among Iran's agricultural outputs. The cultivation of rice spans across various provinces, including Ilam, Lorestan, Zanjan, Gilan, Qazvin, Mazandaran, Golestan, North Khorasan, Razavi Khorasan, Isfahan, Khuzestan, Chaharmahal and Bakhtiari, Kohgiluyeh and Boyer Ahmad, Fars, and Sistan and Baluchistan. Detailed data on the area cultivated, the volumes produced, and the yield of rice across these provinces are provided in Table 2, offering a comprehensive overview of the scope and output of rice agriculture within the nation.

The production of rice in Iran shows that during the studied period, due to increase in the cultivated area, the production of rice has also increased, from 2060.30 thousand tons in 2008 to 4363.44 thousand tons in 2019. Additionally, the yield of rice increased from 3.93 tons per hectare in 2008 to 4.97 tons per hectare in 2019, reflecting a rise of 26% in 12 years. More than 70% of the produced rice belongs to Mazandaran and Guilan provinces. Based on the MAJ report (2021), the mentioned provinces have switched to high-yielding rice cultivars (Shirodi and Fajr cultivars) over the past several years, which has created a tremendous change in rice production in the north of Iran and has an acceptable quality.

The highest yield is seen in Isfahan, Fars, and Mazandaran provinces and the lowest in Qazvin, Zanjan, and Sistan and Baluchistan provinces (Table 2). The remarkable thing in Iran is the great variety of rice cultivars that have high yield. So that in all three mentioned provinces different cultivars have high yield (Rezaei et al. 2021). Mazandaran and Guilan are the largest producers of rice in Iran; they have the largest area under cultivation and account for 70% of the cultivated area and crop production in Iran.

Table 2 The average cultivation area, production, and yield of rice in Iran from 2008 to 2019 (MAJ 2021)

Regions	Province name	(1000 ha) Area	(1000 ton) Production	Yield(ton/ha)
	Mazandaran	217.92	1073.64	4.93
North	Guilan	186.41	758.92	4.07
	Golestan	58.34	272.72	4.68
	Ilam	2.29	10.64	4.66
	Lorestan	4.35	16.67	3.83
	Zanjan	2.51	8.56	3.41
West	Chaharmahal and Bakhtiari	3.24	12.17	3.76
	Kohgiluyeh and Boyer Ahmad	4.80	21.02	4.38
	Sistan and Balochestan	2.16	7.84	3.62
South	Khozestan	63.44	261.83	4.13
	North Khorasan	1.99	7.09	3.55
East	Razavi Khorasan	1.74	6.35	3.65
	Isfahan	7.09	33.74	4.76
Center	Qazvin	2.63	9.47	3.6
	Fars	26.00	128.03	4.93
	Ave	38.99	175.25	4.13
	Max	217.92	1073.64	4.93
	Min	1.74	6.35	3.41

5.2 *Water Footprint Components of Rice in Iran*

In Iran, the average total WF for rice production is quantified at 3037 m^3/ton. A detailed analysis reveals a notable variation across provinces, with Mazandaran exhibiting the lowest WF at 1560 m^3/ton and Zanjan registering the highest at 4950.96 m^3/ton, as illustrated in Fig. 12. Noteworthy is the absence of a green WF in the provinces of Ilam, Lorestan, Zanjan, Chaharmahal and Bakhtiari, Sistan and Baluchistan, Isfahan, and Qazvin, highlighting regional disparities in water resource usage.

North Khorasan leads with the highest green WF at 639.17 m^3/ton, followed by Razavi Khorasan at 441.05 m^3/ton. In terms of blue WF, which encompasses surface and groundwater used in production, Zanjan, Kohgiluyeh and Boyer Ahmad, and Sistan and Baluchistan provinces show the highest figures at 4702.68, 4604.50, and 4322.39 m^3/ton, respectively. Conversely, the northern provinces of Mazandaran, Guilan, and Golestan report the lowest blue WFs, indicating regional variations in water usage patterns.

The analysis of gray WF, which measures the volume of freshwater required to dilute pollutants, identifies Isfahan and Golestan as having the lowest values at 155.76 and 168.94 m^3/ton, respectively, with Khuzestan experiencing the highest gray WF at 270.71 m^3/ton. On average, the gray WF across Iran stands at 208.07 m^3/ton.

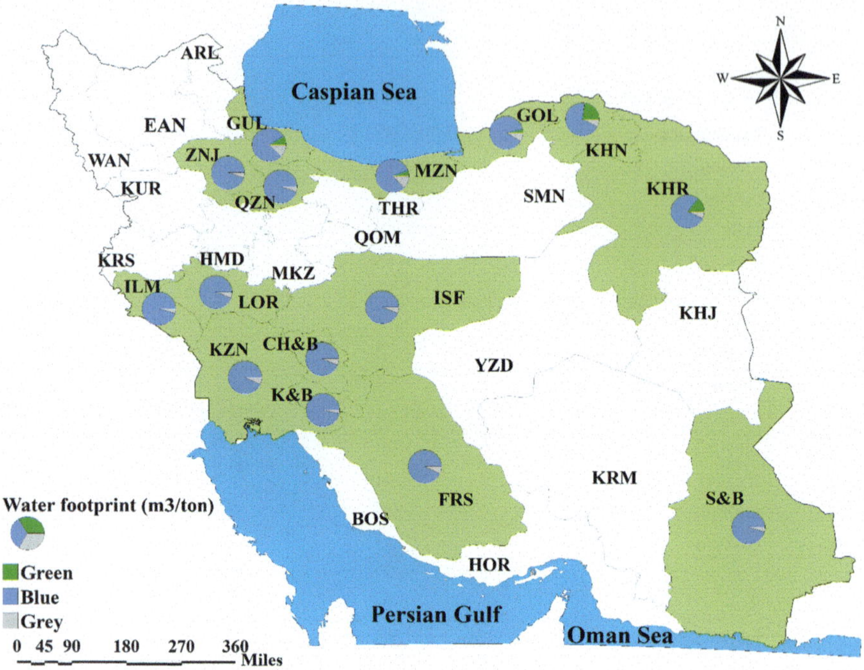

Fig. 12 Average volume of rice water footprint components in the studied provinces of Iran from 2008 to 2019. (Dehghanpir et al. 2024)

This assessment demonstrates that the contributions of blue, green, and gray WFs to the total WF of rice production in Iran are 91.68%, 6.93%, and 1.39%, respectively. Such findings underscore the critical importance of blue water resources in Iranian rice cultivation and highlight the need for sustainable water management practices to address the variability in WF across different provinces.

According to Chapagain and Hoekstra (2011), the average WF of rice on a national scale in Iran was 3100 m^3/ton. In contrast, the present research estimates this value to be 2997 m^3/ton. The most important reason for the difference in the estimated footprint values in the current research is the detailedness of the investigated scale. The difference between water demand, effective precipitation, and crop evaporation–transpiration in each region shows significant changes in the amount of WF, and this situation proves the necessity of using smaller scales for evaluation. It is logical that the data estimated in the plains in each province are more accurate than the estimates on the national scale. On the other hand, various assumptions about input data in WF modeling are also effective on the amount of green WF. Regarding blue water, it should be noted that estimating it with real irrigation data will result in a more accurate estimate of the WF. Therefore, in regional scale studies, estimation of different parameters has higher validity and less uncertainty, despite national scale.

An interesting point to note is the comparison of the WF of rice production in Iran (3037 m^3/ton) with that of India (2357 m^3/ton) (Uma Gowri and Shivakumar 2021), which is the largest producer of rice in the world. With the difference that in India, more than 53% of the WF is green water, while in Iran this amount is only 6.93%. As mentioned earlier, more than 80% of rice production and cultivated area in Iran belong to the north (Golestan, Guilan and Mazandaran), which has the highest rainfall (600–2100 mm) in Iran. Although the proportion of precipitation during the rice growth period is much lower than the entire year, changing the cultivation period to maximize the use of effective precipitation can effectively increase the share of green water compared to other components of the WF.

The highest share of WF (91%) in rice production is related to blue water. The use of drought-resistant genotypes, the use of modern irrigation methods called rice drying, which is currently used instead of flooded rice cultivation, can be a suitable way to reduce the WF of rice. Mainly, this method reduces the deep and lateral penetration of water in the soil and reduces evaporation (Shekhawat et al. 2020). This method has also been piloted in Khuzestan (Gilani et al. 2019) and Isfahan (Ramazani and Dehghani 2021).

5.3 AWSI and RWSI in Iran

RWSI is calculated by assessing the ratio of the total WF in rice production to the AWR within a province. Consequently, provinces with lesser access to water resources exhibit higher values of WSI. Figure 13 illustrates the spatial variation of RWSI across rice-producing provinces, juxtaposed with the AWSI. According to the findings, RWSI values vary significantly, from as low as 0.16 in Mazandaran province to as high as 1.32 in Sistan and Baluchistan province.

This study indicates that AWSI exceeds 0.6 in Lorestan, Zanjan, Qazvin, North Khorasan, Razavi Khorasan, Isfahan, Khuzestan, Fars, and Sistan and Baluchistan provinces, signaling exceptionally high water stress levels in the agricultural sector within these regions. In contrast, Mazandaran, Guilan, Golestan, Fars, and Khuzestan provinces exhibit AWSI values below 0.3, denoting lower levels of water stress. This disparity highlights the varying degrees of water stress challenges faced by different provinces in the production of rice and other agricultural products.

Overall, the average RWSI throughout the study period stands at 0.5, positioning Iran's rice production within the medium-to-high water stress category, as per the classification provided in Table 2. This categorization underscores the pervasive normal of water stress in rice production across the country, emphasizing the need for strategic water management and conservation practices to mitigate these challenges.

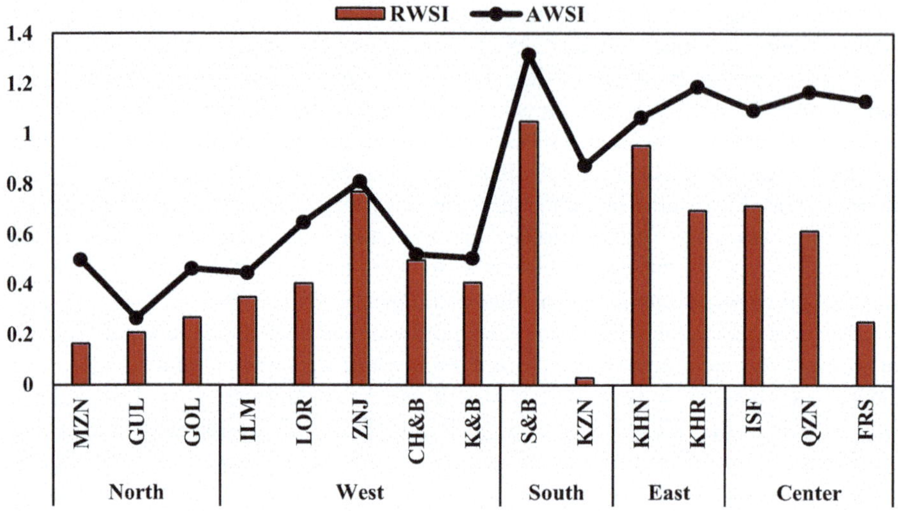

Fig. 13 Comparison of AWSI and RWSI at the provincial scale from 2008 to 2019. (Ref: Authors)

5.4 Temporal Variation of AWSI and RWSI in Iran

RWSI consistently registers lower than AWSI across the observed periods. This difference arises because only a portion of Iran's total water resources are allocated for rice production in the examined provinces, with the remainder utilized for other agricultural products. The analysis of temporal variations between these two indices does not reveal any significant trend, indicating a stable pattern over the years.

The findings demonstrate that the average AWSI within the agricultural sector across the investigated provinces stands at 0.92. This high value underscores the intense level of water stress experienced in these regions, driven by agricultural demands. When compared to the RWSI in Iran, which is calculated at 0.5, it becomes evident that other agricultural products account for approximately 42% of the water stress on the nation's water resources, as illustrated in Fig. 14. This discrepancy highlights the broader impact of agricultural practices beyond rice production on water stress levels, suggesting the need for comprehensive water management strategies that encompass all agricultural activities to mitigate water scarcity issues effectively.

5.5 Water Deprivation in Rice Production at the Provincial Scale

The analysis of data from 2008 to 2019 reveals that, on average, rice production in the studied provinces results in a total water deprivation of 1073 MCM. Notably, the

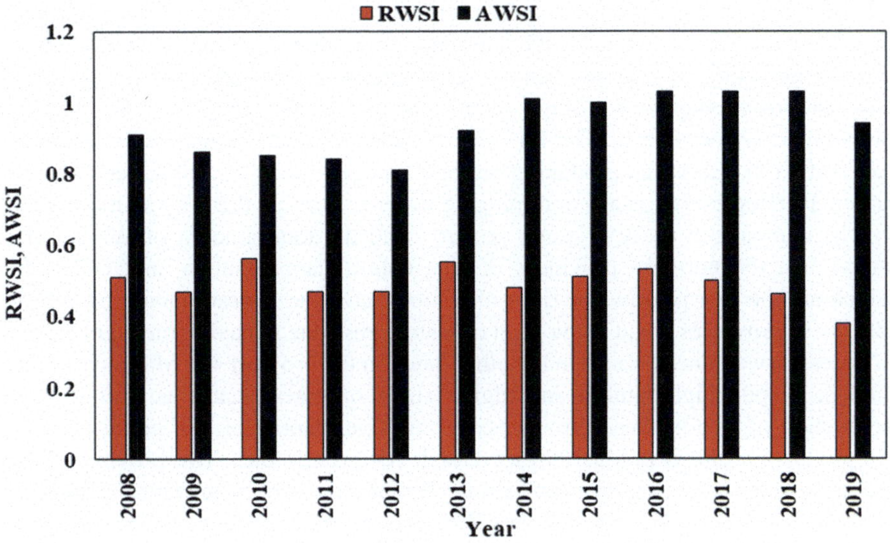

Fig. 14 Temporal variation of AWSI and RWSI in Iran from 2008 to 2019. (Ref: Authors)

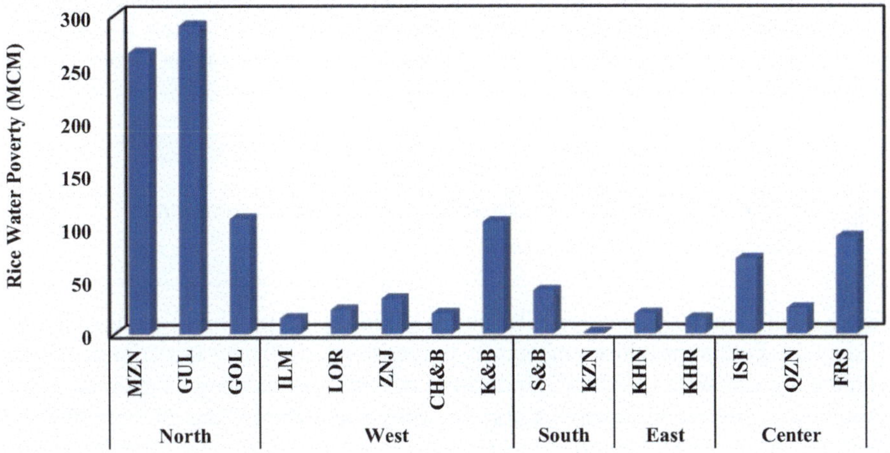

Fig. 15 Water deprivation in rice production at the provincial scale in Iran from 2008 to 2019. (Ref: Authors)

province of Khuzestan experiences the lowest level of water deprivation, recording just 11 MCM, closely followed by Ilam and Razavi Khorasan with 14.65 MCM and 14.74 MCM, respectively. On the other end of the spectrum, Gilan and Mazandaran provinces face the most significant water deprivation, with figures reaching 290.24 and 264.90 MCM, respectively, as detailed in Fig. 15.

This disparity highlights a distinct regional pattern within Iran, where northern provinces such as Mazandaran, Guilan, and Golestan endure the highest levels of water deprivation in rice production. The extensive cultivation of rice in these northern areas, coupled with high production and yield rates, contributes to their pronounced water poverty. The findings underscore the challenges posed by water deprivation in Iran's rice-producing regions, particularly in the north, necessitating targeted strategies to manage and mitigate water scarcity in these agricultural hubs.

The comparison of WSI, water poverty, and the total amount of WF for rice shows that the northern provinces of Iran (Mazandaran, Guilan, and Golestan), which are the first priority for rice cultivation, have the lowest amount of WF. In fact, these provinces are at a low level of water stress due to having the highest level of access to water resources. On the other hand, in the provinces of Mazandaran and Guilan, despite high rainfall and higher share of green water, as well as high production of rice in these provinces, the average amount of WF components in the production of this product has increased. As a result, these two provinces have the most water poverty in rice production. The provinces of Sistan and Baluchistan, Isfahan, North Khorasan, Qazvin, and Zanjan have the highest total WF, especially the blue WF, in rice production. Due to limited access to water resources and less rainfall, they are at a high water stress level in terms of WSI. In general, the most water poverty is related to the provinces (Mazandaran, Guilan, and Golestan) that produce the most rice. The results of Cao et al. (2017) in China also confirm that water poverty also increases with the increase of access to water, because the motivation to increase production and the volume of the footprint due to the increase in production also increase.

5.6 Water Self-Sufficiency Index and Water Dependency Index in Rice Production

Figure 16 illustrates the average WSS and WD indices for provinces engaged in rice production. Guilan province emerges as the most self-sufficient in rice production, with the highest WSS of 89.54%, indicating minimal dependency on external water resources. Conversely, Razavi Khorasan province exhibits the highest WD at 94.32%, underscoring a complete reliance on water resources from other provinces for rice cultivation. These metrics reveal a significant variance in water resource management and dependency across rice-producing regions, highlighting Guilan's capability to meet its rice production water needs internally, while Razavi Khorasan's rice agriculture depends heavily on external water supplies.

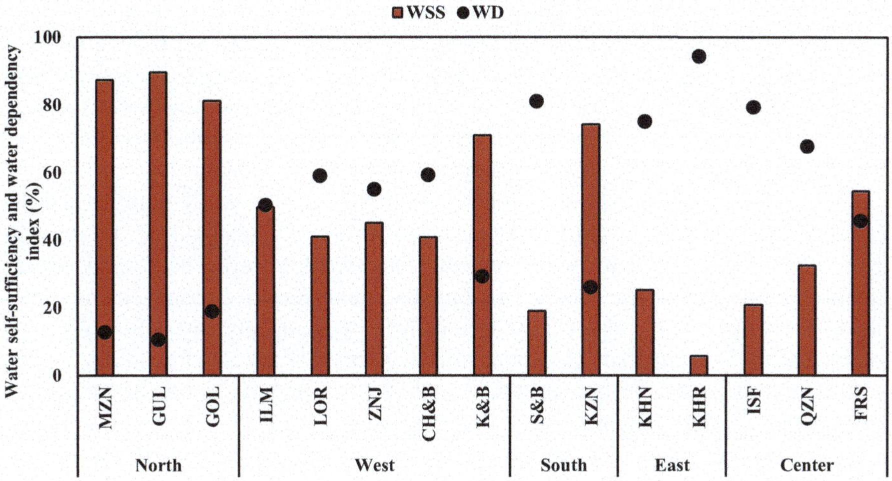

Fig. 16 Water self-sufficiency and water dependency in rice production at the provincial scale in Iran from 2008 to 2019. (Ref: Authors)

6 Conclusion

The findings of this study highlight a significant reliance on blue WFs over green WFs in the production of wheat and rice in Iran, underscoring the pivotal role of blue water resources in the agricultural process. This distinction illustrates a substantial consumption of surface and underground water resources, as opposed to reliance on green water resources, such as rainwater. The comparatively low green WF indicates that rainfall in Iran is insufficient to support the cultivation of wheat and rice, emphasizing the critical importance of water management in the agricultural sector. This scenario underscores the considerable impact that wheat and rice production have on Iran's water resources, affirming that the implications of these agricultural activities on water availability and sustainability cannot be overlooked.

The findungs of this study suggest that the variations in the WSI, water poverty, and WF, which are crucial for assessing water scarcity in wheat and rice production, are predominantly influenced by the utilization of blue water resources over green water resources. Hence, incorporating assessments of water deprivation alongside the WSI and WF offers a robust framework for evaluating water scarcity in relation to agricultural production in Iran. The substantial WF observed in wheat production within the provinces of Fars, Golestan, Kurdistan, Razavi Khorasan, Hamadan, Khuzestan, Kermanshah, and Ardabil have escalated water poverty levels. Similarly, rice production in Kohgiluyeh and Boyer Ahmad, Sistan and Balochistan, Chaharmahal and Bakhtiari, Zanjan, and Khuzestan provinces exhibit high WFs, leading to increased water deprivation. Provinces exhibiting lower levels of water deprivation and stress, and consequently a smaller WF, may possess a greater potential for impactful resource management strategies. WSI considers only blue water consumption, while WWSI and RWSI provide a more comprehensive assessment by including both blue and green water consumption, as well as gray water consumption needed to dilute chemical fertilizers in wheat production. This makes WWSI and RWSI more accurate indicators for evaluating water scarcity and management in the production of wheat and rice. Given the prominence of the blue WF relative to green and gray footprints, a significant factor contributing to water stress in these crops is the high consumption of blue water resources.

This analysis underscores the critical challenge posed by water scarcity and the resulting stress on food security and the sustainability of wheat and rice production in Iran. The implications for water resources and water security necessitate urgent attention, emphasizing the need for sustainable water resource management strategies in grain production. Providing these insights to managers and policymakers is crucial for identifying and addressing areas vulnerable to water scarcity. Future research should extend beyond wheat and rice to include a broader range of agricultural products, enabling a comprehensive evaluation of the environmental impacts associated with WU in agriculture and formulating recommendations to alleviate water pressure in the sector.

References

Abdollahzadeh Kahrizi R, Moghaddam AHK, Merufinia E (2022) Investigating virtual water and agricultural water productivity index in crops of Poldasht plain. Soil and Water Modeling and Management 3(1):54–68

Allan JA (1998) Virtual water: a strategic resource global solutions to regional deficits. Groundwater 36(4):545–546

Bazarfshan O, Yahyazadeh M, Jamshidi S, Zamani H (2022) Spatial prioritization of tomato cultivation based on water footprint, land productivity, and economic indices. Irrig Drain 71(5):1363–1378

Cao XC, Wu PT, Wang YB, Zhao XN (2014) Assessing blue and green water utilisation in wheat production of China from the perspectives of water footprint and total water use. Hydrol Earth Syst Sci 18(8):3165–3178

Cao XC, Mengyang W, Xiangping G, Yalian Z, Yan G, Nan W, Weiguang W (2017) Assessing water scarcity in agricultural production system based on the generalized water resources and water footprint framework. Sci Total Environ 609:587–597

Cao X, Huang X, Huang H, Liu J, Guo X, Wang W, She D (2018) Changes and driving mechanism of water footprint scarcity in crop production: a study of Jiangsu Province, China. Ecol Indic 95: 444–454

Chapagain AK, Hoekstra AY (2011) The blue, green and grey water footprint of rice from production and consumption perspectives. Ecol Econ 70(4):749–758

Dehghanpir S, Bazrafshan O, Ramezani Etedali H, Holisaz A, Ababaei B (2022) Evaluation of water shortage in wheat production at Iran. Iran J Ecohydrol 9(4):719–732. https://doi.org/10.22059/ije.2023.348718.1683

Dehghanpir S, Bazrafshan O, Ramezani Etedali H, Holisaz A, Collins B (2023) Water scarcity assessment in Iran's agricultural sector using the water footprint concept. Environ Dev Sustain:1–26

Dehghanpir SH, Bazrafshan O, Ramezani Etedali H, Holisaz A, Ababaei B (2024) Evaluation of water stress index and water poverty in rice production based on the water footprint concept in Iran. Water Soil Manage Model 4(1):18–35

Ebrahimi Mahmoudi H, Pishvaei MS, Teymouri E (2021) A two-stage model for rice cultivation preparation considering dynamic uncertainty: a case study in Iran. Indus Manage Perspect 11(42):145–176. https://doi.org/10.52547/jimp.11.2.145

Elliott J, Deryng D, Müller C, Frieler K, Konzmann M, Gerten D, Wisser D (2014) Constraints and potentials of future irrigation water availability on agricultural production under climate change. Proc Natl Acad Sci 111(9):3239–3244

Falkenmark M (2001) The greatest water problem: the inability to link environmental security, water security and food security. Int J Water Resour Dev 17(4):539–554

Fathian M, Bazrafshan O, Jamshidi S, Jafari L (2023) Impacts of climate change on water footprint components of rainfed and irrigated wheat in a semi-arid environment. Environ Monit Assess 195(2):324

Fu H, Chen Y, Yang X, Di J, Xu M, Zhang B (2019) Water resource potential for large-scale sweet sorghum production as bioenergy feedstock in northern China. Sci Total Environ 653:758–764

Gerbens-Leenes PW, Mekonnen MM, Hoekstra AY (2012) The water footprint of poultry, pork and beef: a comparative study in different countries and production systems. Water Resour Indus 1:25–36

Gilani A, Absalan S, Jalali S, Behbahani L (2019) The effect of sprinkler irrigation on grain yield, yield components and water use efficiency of Rice cultivars under drill-seed cultivation in Khuzestan. Irrig Sci Eng 42(2):63–73. https://doi.org/10.22055/jise.2017.19659.1409

Hadi M, Jalili M, Heris A (2017) Assessing the wheat yield under irrigated and rainfed farming and evaluating the possibility of supplemental irrigation of rainfed by water stored in deficit irrigated farming abstract. Iran J Irrig Drain 3(11):403–411

Hoekstra AY, Hung PQ (2002) Virtual water trade: a quantifcation of virtual water flows between nations in relation to international crop trade. Value of water research report series no. 11. UNESCO-IHE, Delft, The Netherlands

Hoekstra AY, Mekonnen MM (2012) The water footprint of humanity. Proc Natl Acad Sci 109(9): 3232–3237

Hoekstra AY, Chapagain AK, Aldaya MM, Mekonnen MM (2011) The water footprint assessment manual: setting the global standard. Routledge

Jamshidi S, Zand-Parsa S, Naghdyzadegan Jahromi M, Niyogi D (2019a) Application of a simple Landsat-MODIS fusion model to estimate evapotranspiration over a heterogeneous sparse vegetation region. Remote Sens 11(7):741

Jamshidi S, Zand-parsa S, Pakparvar M, Niyogi D (2019b) Evaluation of evapotranspiration over a semiarid region using multiresolution data sources. J Hydrometeorol 20(5):947–964

Jamshidi S, Murgia T, Morales-Ona AG, Cerioli T, Famoso AN, Cammarano D, Wang DR (2023) Modeling interactions of planting date and phenology in Louisiana rice under current and future climate conditions. Crop Sci 64(4):2274–2287

Liu J, Yang H, Gosling SN, Kummu M, Flörke M, Pfister S, Oki T (2017) Water scarcity assessments in the past, present, and future. Earth's Future 5(6):545–559

Madani K (2014) Water management in Iran: what is causing the looming crisis?. J Environ Stud Sci 4:315–328

MAJ (2021) Iran Agriculture Bulletin. Ministry of Agriculture Jihad, Agriculture Jihad Press, Tehran (InPersian)

Pfister S, Koehler A, Hellweg S (2009) Assessing the environmental impacts of freshwater consumption in LCA. Environ Sci Technol 43(11):4098–4104

Rafiei Sardooi E, Bazrafshan O, Jamshidi S (2024) Modeling the water security in a watershed using the water footprint concept and water scarcity indicators. Water Supply 24(1):235–253

Ramazani A, Dehghani M (2021) Application of trickle irrigation (T-tape) in dry direct-seeded Rice (case study of Lenjan region of Isfahan). Iran Water Res J 15(2):119–127

Raskin P, Gleick P, Kirshen P, Pontius G, Strzepek K (1997) Water futures: assessment of long-range patterns and problems. Comprehensive assessment of the freshwater resources of the world. SEI, Stockholm

Rezaei G, Khaledian M, Kavoosi-Kalashami M, Rezaei M (2021) Comparison of water productivity indices and virtual water in major rice producing provinces in Iran. Iran J Irrig Drain 15(3): 634–644

Shekhawat K, Rathore SS, Chauhan BS (2020) Weed management in dry direct-seeded rice: a review on challenges and opportunities for sustainable rice production. Agronomy 10(9):1264

Sullivan C (2002) Calculating a water poverty index. World Dev 30(7):1195–1210

Uma Gowri M, Shivakumar KM (2021) India rice export and virtual water trade. J Appl Nat Sci 13 (SI):43–46

Wang YB, Wu PT, Engel BA, Sun SK (2015) Comparison of volumetric and stressweighted water footprint of grain products in China. Ecol Indic 48:324–333. https://doi.org/10.1016/j.ecolind.2014.08.014

Water Footprint (WFP) in Hydropower Plants Using Observed Meteorological Datasets: The Case of the Funil Hydropower Plant, Southeast Brazil

Eduardo de O. Bueno, Carlos R. Mello, Jorge A. Guzman, and Geovane J. Alves

1 Introduction

At the beginning of the twenty-first century, the concept of water footprint (WFP) was introduced as a sustainable water indicator that allows the analysis of the present, past, and future economic performance of the total volume of water consumed, directly or indirectly, during the production of one unit of a product or service. The term was chosen as an analogy to the ecological footprint (Hoekstra et al. 2011). The quantification of the WFP is an analytical process that accounts for the total water consumption throughout the production chain, where consumption

E. O. Bueno
Water Resources Department, School of Engineering, Federal University of Lavras, Lavras, MG, Brazil

C. R. Mello (✉)
Water Resources Department, School of Engineering, Federal University of Lavras, Lavras, MG, Brazil

Department of Agricultural and Biological Engineering, College of ACES, University of Illinois at Urbana-Champaign, Urbana, IL, USA
e-mail: crmello@ufla.br; cdemello@illinois.edu

J. A. Guzman
Department of Agricultural and Biological Engineering, College of ACES, University of Illinois at Urbana-Champaign, Urbana, IL, USA
e-mail: jag@illinois.edu

G. J. Alves
Hidrológica Soluções, Lavras, MG, Brazil

Centro Universitáro de Lavras – UNILAVRAS, Lavras, MG, Brazil
e-mail: geovane@hidrologicasolucoes.com.br

S. S. Muthu (ed.), *Sustainability and Water Footprint*, Environmental Footprints and Eco-design of Products and Processes, https://doi.org/10.1007/978-3-031-70810-7_4

Table 1 Spatial and temporal resolutions in WFPP calculation (Hoekstra et al. 2011)

Spatial resolution	Temporal resolution	Data source	Use of the results
Global average (Level A)	Annual	Global datasets about the consumption and pollution of water by product or process are available.	Awareness-raising actions, preliminary identification of the components contributing most to the total water footprint, and development of global water consumption projections.
National, regional, or watershed (Level B)	Annual or monthly	Like level A, using more regional datasets.	Preliminary identification of spatial dispersion and variability; support for identification of critical areas and decisions on water allocation.
Watershed (Level C)	Monthly or daily	Empirical or measured datasets (an improved database with local data throughout the year)	Support the sustainability assessment of a water footprint and formulate a strategy to reduce the footprint and the associated local impacts.

refers to the appropriation of water for a specific purpose in the same place and time used to define the quantification of WFP. Therefore, WFP measures water consumption for a specific purpose, which will no longer be available in the same place and time for another purpose (Hoekstra 2017), and the higher the WFP, the less efficient is the water use in producing goods (Bueno et al. 2016).

WFP can be estimated individually per person, product, industrial plant, country, watershed, or geographical region on a monthly or annual basis (Hoekstra et al. 2011), the lowest level of detail (level A), commonly sufficient and suitable for awareness-raising purposes. On level B, WFP is assessed based on the national or regional average or specific watersheds with typical applications on decision-making processes regarding water allocation (water resource management). Finally, at level C, WFP estimates are geographically and temporally explicit, based on precise and detailed information on water consumption and pollution, preferably with field data with a minimum spatial resolution on a minimum monthly basis temporal resolution. This is the case of the FPP presented in this chapter. Besides, specific WFPs per product development include a seasonal analysis of intra-annual variations for the highest spatial and temporal level of detail suitable for formulating specific local WFP reduction strategies. In Table 1, further details were incorporated regarding the details levels described above.

1.1 Hydropower Generation in Brazil and Water Footprint

Brazil ranks third in the world in terms of hydroelectricity generation, behind Canada and China (International Energy Agency—IEA 2017). Due to water scarcity, more significant fossil fuel energy generation has increased in the Brazilian electricity matrix in recent years, with a reduction of the hydroelectricity plants (IEA 2017). Despite this aspect, the country's energy matrix is predominantly based on hydroelectricity generation, with 1265 hydropower plants (HPPs) in operation, with different strategies (e.g., micro- and small plants, run-of-river, and large and accumulation reservoirs), reservoir size, and power installed, which corresponds to 69% of the total electricity generation in the country (Agência Nacional de Energia Elétrica—ANEEL 2017).

In the current scenario of climate change and scarcity of water resources, a growing concern regarding the sustainability of HPPs has been observed, leading to more detailed studies on the relationship between water use and hydroelectricity (Liu et al. 2015). Although HPPs are renewable energies, i.e., they are considered clean energy sources, recently, there has been growing concern about their environmental sustainability, including, for example, the large amount of water lost to evaporation from the reservoirs (Hoekstra et al. 2011). Hydroelectricity is an energy source classified as having a low environmental impact, mainly due to its renewable nature and low greenhouse gas emissions. However, water consumption is one of the most significant per unit of energy produced among the renewable and non-renewable energy sources (Zhao and Liu 2015).

In recent years, there has been a relevant and intense debate about whether the use of water by HPPs is, in fact, non-consumptive (Aguilar et al. 2011; Bueno et al. 2016; Mekonnen and Hoekstra 2011). This issue is increasingly resonating due to the concept of "WFP", which consists of the water consumption required to produce a product unit, considering all the stages of its production chain (Hoekstra et al. 2011). The approach is more specific for HPPs, as the WFP is calculated in terms of the ratio of evaporation from the reservoirs and the energy produced in a given period, expressed in $m^3.GJ^{-1}$.

The Intergovernmental Panel on Climate Change (IPCC) has intensified the search for standardized methodologies for estimating water losses due to various forms of energy generation so that a rigorous comparison between them can be conducted (Bakken et al. 2013; IPCC 2012). In this sense, the WFP has been an essential indicator for analyzing the sustainable and efficient use of water resources for hydroelectricity. The WFP of Brazilian HPPs can be linked to the National Interconnected System (SIN) operation carried out by the National Electricity System Operator (ONS), which is responsible for defining the amount of energy generated at each plant. In Brazil, the hydroelectricity system is entirely connected, which requires an operator system for the facilities. In addition, the operation of the reservoirs can also impact the WFP behavior, i.e., the system works synergically, and a given plant can have other functions than energy production, like a hydrological function, when a cascade of plants is established.

HPPs are characterized by having a high WFP. However, WFP estimates based on an inadequate methodological approach and average data can lead to inconsistent results, implying erroneous and biased conclusions. Hence, what are the electricity companies' concerns? Two of the major concerns are: (1) impacts on public view (and share market) of products that use energy from hydropower generation as an input in their production process, especially in the international market; and (2) induction of policies to encourage the use of other sources of energy generation with a lower WFP, such as non-renewable sources with more significant environmental impacts.

Therefore, in this context, we studied a methodological approach for WFP calculation in HPPs based on local weather observations and incorporation of the thermodynamics of the reservoirs by monitoring the water temperature. We used the Funil HPP in the southern Minas Gerais state, Brazil, as a case study.

2 Concepts

2.1 Reservoir Evaporation

This section presents the most used methods for reservoir evaporation estimates, based on several worldwide studies. The methods for estimating evaporation from reservoirs are classified as mass transfer (aerodynamic), energy balance, combined methods, water balance method, empirical equations, and direct measurements (Bueno et al. 2020).

In the first approach, evaporation is based on water vapor transport by a turbulent diffusion process. Then, based on the energy balance, evaporation consumes incident radiation. Penman (1948) combined the evaporative power of the air (e.g., temperature, humidity, wind speed) with solar radiation, giving rise to one of the most widely used methods for evaporation estimates. Using the water balance method to estimate reservoir evaporation is theoretically accurate, based on the mass conservation principle. The problem is that many technical issues are found in measuring the various components of the water balance, constraining the use of this method (Bueno et al. 2016; McMahon et al. 2013). The sensitivity and accuracy of the measurement can imply significant errors in evaporation estimates. However, in addition to the classic instruments, modern high-frequency detection sensors (eddy covariance, for example) have been used in recent years to measure the water vapor flux from a liquid surface to the atmosphere (Dias et al. 2002; Rosenberry 2007; Tanny et al. 2008, 2011).

Some experiences in evaporation estimates can be highlighted. Let us start with the Abtew (2001) study. A comparative study used seven methods for estimating evaporation from reservoirs in South Florida, USA, including water balance, Class A tank, mass transfer, and Penman and Priestley–Taylor equations. This study pointed out that the mass transfer method presents greater errors than the others in estimating evaporation. The Priestley–Taylor and Penman methods overestimated evaporation compared to the other methods. The water balance method also overestimated evaporation by approximately 10% compared to the other methods, which is explained by infiltration losses in the reservoirs.

The Rosenberry (2007) study is also relevant due to the number of tested methods. In this evaporation study in the USA, carried out in New Hampshire, the performance of 15 methods was evaluated, considering the Bowen ratio energy balance (BREB) method as the reference for comparisons. Several methods had significant errors compared to the values obtained with the BREB method. The results of the Penman and Priestley–Taylor methods were the closest to the BREB results and could be recommended.

An interesting investigation was conducted in Egypt, which can also be highlighted because of the location and the number of methods. In this case, a sensitivity study was developed considering 12 evaporation estimation methods using data collected at a floating climatological station in the Aswan High Reservoir in southern Egypt. Among the main conclusions, the sensitivity indices revealed that the energy term in the BREB method affects evaporation estimates the most. The most sensitive parameter for the combined methods (Penman and Penman–Monteith) is the variation in heat stored in the reservoir. For the other methods, the most sensitive parameters were solar radiation and air temperature. The results indicate that many parameters of the models can be fixed without significantly affecting the evaporation estimates (Elsawwaf et al. 2010).

Finally, McMahon et al. (2013) presented an extensive literature review on the methods for estimating evaporation from shallow and deep reservoirs, covering conceptual aspects, limitations, and methodological guidelines, including application examples. The following methods were highlighted among the most frequently applied ones in reservoir evaporation studies: the Penman method, Penman–Monteith, Priestley–Taylor, Class A tank, water balance, and Morton's CRLE model (1986).

2.1.1 The Penman–Monteith Method for Evaporation Estimates from Reservoirs

Among the methods derived from Penman (1948), the Penman–Monteith method (Monteith 1965) and its various later adaptations stand out for considering the variation in the heat stored in the reservoir when estimating evaporation. In addition, the Penman–Monteith method adapted to evaporation estimation does not assume that the water temperature is the same as the air temperature, i.e., monitoring the water temperature over time is important. In these models, when the water temperature is not monitored, it is calculated indirectly (Bueno et al. 2016). Studies like McJannet et al. (2008) used sensitivity analysis to assess errors in estimating evaporation from reservoirs in the Murray–Darling basin, Australia, by varying the input data in the proposed Penman–Monteith equation. The results are summarized in Table 2. The sensitivity in estimating evaporation using the Penman–Monteith method is greater for errors caused by determining the reservoirs' area. The larger the surface water area, the lower the aerodynamic function and, consequently, the lower the daily evaporation rate. However, the most significant influence of the reservoir's area error occurs directly in the error in estimating the water volume evaporated since it is obtained by the product of the reservoir's area and the evaporation depth. Thus, for large reservoirs with volume (level) varying over time, it is fundamental to

Table 2 Penman–Monteith sensitivity method for reservoir evaporation

Variable	Range of the values	Variation on evaporation estimate
Temperature	±2 °C	±3%
Solar radiation	±10%	±6%
Wind velocity	±50%	±7%
Elevation	±50%	±1%
Latitude	±2°	±1%
Depth	±50%	±1%
Reservoir's area	±20%	±20%

Source: McMahon et al. (2013)

access the depth–area–volume curves and their change due to transformations in the reservoirs, i.e., updated curves.

2.1.2 Heat Storage Variation on the Reservoirs

The importance of considering the variation in the heat stored in the reservoirs when estimating evaporation has been assessed in many studies. Thus, it can be inferred that the influence of this process varies depending on (i) the seasonality and intensity of the incident solar radiation; (ii) the reservoir's depth; (iii) the time step used in the calculation; and (iv) the accuracy of the measurement or indirect calculation of the water temperature (Bueno et al. 2016; Dias and Kan 2008; Finch 2001; McJannet et al. 2008; McMahon et al. 2013). For deeper reservoirs (>2 m), on a monthly scale, it is necessary to consider the effect of the variation in thermal energy storage in the reservoir on the evaporation estimate.

In the long term, the variation in heat stored in the reservoir tends to be close to zero because the heat gained during summer, when the reservoir is already warmed, is lost during winter, when it cools down (Dias and Rocha 1999). Hence, the monthly evaporation cycle in deep reservoirs lags the incident solar radiation cycle by a few months. This is due to the absorption (or release) of energy to heat (or cool) the water in the reservoir, i.e., the variation in the heat stored in the reservoir. The deeper the reservoir, the greater the lag between the monthly estimate of evaporation and the incident solar radiation. As the reservoir gets deeper, the maximum evaporation values occur far from the months when the incident solar radiation is at its highest (Dias and Kan 2008). As an example, a study compared evaporation estimates from the Baratz Reservoir in the Mediterranean (Sardinia, Italy), obtained using the Penman (1948) and BREB methods (Giadrossich et al. 2015). The heat-stored variation in the reservoir was obtained based on the vertical profile of the reservoir's water temperature. On an annual scale, evaporation estimated by the Penman method, which disregards the heat-stored variation in the reservoir, was 18% higher than the BREB method, which includes heat-stored variation, with monthly differences ranging from −38% to 60%.

2.1.3 Evapotranspiration

The main methods for estimating evapotranspiration can be classified as water balance, combined method (Penman, Penman–Monteith, Penman–Monteith–FAO), empirical methods (Priestley–Taylor, Linacre, Thornthwaite), and direct (lysimeters, water balance) and indirect (evaporimeter tank, Piché evaporimeter) measurements. Many of these methods vary due to adjustment and calibration of local parameters, increasing the number of available methods (Allen et al. 1998).

There are several methods for estimating potential evapotranspiration (ETP), ranging from the simplest, based purely on empiricism, to those with a consistent physical basis. ETP is "the amount of water transpired in a given time by a short green crop, completely shading the ground, of uniform height and with adequate water status in the soil profile" (Penman 1948). Basically, it means there is no water availability restriction, like a reservoir's surface. This diversity of methods is due to the complexity of the water transfer from the soil–plant system to the atmosphere, the different climate features, and the difficulty in obtaining the meteorological elements needed to feed these methods. Thus, it is necessary to determine which method best suits local conditions, data availability, and the scale (spatial and temporal) required for the expected results.

It is interesting to consult the McMahon et al. (2013) study, which also presented an extensive literature review on evapotranspiration estimation methods, covering conceptual aspects, limitations, and methodological guidelines. Among the methods most frequently applied in evapotranspiration studies, the following were highlighted: the Penman–Monteith method; FAO-56 reference evapotranspiration; the Priestley–Taylor and Thornthwaite empirical equations; the water balance; and Morton's CRAE model (1986).

2.1.4 Net Evaporation from Reservoirs

A fundamental concept for the WFP of HPP projects is net evaporation, which consists of the difference between reservoir evaporation and the actual evapotranspiration of the reservoir's area before its implementation. For that calculation, it is necessary to adopt a Kc coefficient (crop coefficient), which represents the ratio between actual evapotranspiration (ETa) and ETP. The Kc coefficient varies according to plant and soil moisture conditions, varying theoretically from 0 to 1; the unit value means ETa= ETP, i.e., there is no water restriction.

One of the methods widely used to estimate reservoir evaporation and evapotranspiration in the area before filling the reservoir consists of the complementary relationship areal evapotranspiration (CRAE) and complementary relationship reservoir evaporation (CRLE) models, proposed by Morton (1983a, b). Later, these models were compiled into a single computer program, the WREVAP—Operational Estimates of Areal Evapotranspiration and Reservoir Evaporation (Morton et al. 1985), widely used when evaporation is studied large scales using climate reference values (Bueno et al. 2020). Most of the WFP studies in HPPs have been conducted following this procedure.

3 Water Footprint (WFP) in Hydropower Plants

3.1 WFP in the Hydroelectricity Generation

WFP investigations of hydroelectricity generation began with Gerbens-Leenes et al. (2009) (Table 3). Several studies on WFP of different energy sources (fossil and renewable energy) have been conducted (Hoekstra 2017; Mekonnen et al. 2015, 2016; Wilson et al. 2012).

Hydroelectricity plants are characterized by having a high WFP compared to other sources of electricity generation, except biomass. However, several criticisms of the generalized adoption of a global average value for this indicator have been made, like methodological issues, climate, hydrological, constructive, and technological specificities of each plant and its location (Bakken et al. 2013; Herath et al. 2011)

The WFP of photovoltaics and wind sources is two orders of magnitude smaller than that of fossil fuels. In comparison, the WFP of bioenergy and hydroelectricity is two to three orders of magnitude larger than that of fossil fuels and nuclear energy (Hoekstra 2017). The WFP of thermoelectric power plants could be higher due to the high percentage of water recirculation and reuse in closed-loop processes. The WFP of biomass power plants is characterized by high water consumption associated with producing raw materials (Mekonnen et al. 2016). WFP depends on the crop, agricultural production system, and climate for biomass power plants. In this case,

Table 3 WFP of the primary source of electricity production

Electricity primary source		Global water footprint ($m^3.GJ^{-1}$)
Not renewable	Nuclear	0.09
	Natural Gas	0.11
	Coal	0.16
	Petroleum	1.06
Renewable	Eolic energy	0.001
	Solar energy	0.27
	Hydraulic energy	22.3
	Biomass energy	71.5

most water consumption is linked to the crop evapotranspiration used as fuel, whether from rain or irrigation

According to different sources, hydroelectricity has shown higher WFP of electricity production in the US than other renewable and non-renewable sources (Wilson et al. 2012). In this case, the WFP of different sources was considered, justifying their findings that water stored in reservoirs, as well as that used for cooling in thermoelectric plants, is affected by rising temperatures, reducing its quality and causing losses in aquatic biodiversity

3.2 WFP of Hydropower Plants

In recent years, WFP of hydroelectricity has been a hot spot, prompted by the question of whether hydroelectric plants are only users installed in the river (non-consumptive) or if they consume water (consumptive) (Mekonnen and Hoekstra 2011). Despite these concerns, the WFP of HPPs is commonly categorized in the blue WFP category. Hence, this indicator is calculated by the ratio between plant water consumption and energy produced in a given HPP and time, usually expressed in $m^3.GJ^{-1}$. Its behavior is specific to each plant as a function of factors affecting energy production and evaporation from the reservoirs:

- installed power and turbine efficiency;
- demand for electricity;
- reservoir operating guidelines;
- topography (headwater);
- hydrological aspects (rainfall patterns, natural river flows);
- reservoir size (evaporating surface),
- climatic conditions (air temperature, solar radiation); and,
- nearby land use.

The WFP of HPPs is proportional to evaporation. Thus, it depends on the reservoir's area and meteorological behavior, mainly solar radiation and vapor pressure deficit directly influencing evaporation. On the other hand, WFP is inversely proportional to the electric energy produced on the HPP, i.e., it is linked to the installed power and the efficiency of the turbines. In addition, the topographical condition of the dam site also influences it, and designing a headwater that can maximize the generation is highly strategic for considering a smaller reservoir, which would reduce WFP and the efficiency of the water uses. Thus, in this scenario, the WFP will be lower. The rainfall regime also influences WFP since more rainfall in the watershed leads to higher inflows and guarantees optimum operating levels for the reservoirs and, consequently, a lower WFP. Similarly, the size of the river on which the HPP is installed also impacts the WFP, with higher natural flows enabling more significant energy production without the need for large reservoirs. This results in lower evaporation rates and, therefore, a lower WFP.

In recent years, the construction of run-of-river HPPs in Brazil has been a trend to reduce socio-environmental impacts. This type of HPP accounts for small reservoirs with reduced environmental impacts like deforestation. The ratio between the reservoir's area and the installed power in the HPP (ha.MW^{-1}) has been linear, with increasing correlation with the WFP. Large reservoirs with low installed power have a larger WFP and non-sustainable water use. In Brazil, the first HPPs have large reservoirs with low installed power, leading to high WFP, mainly those in the northeast region (Bueno et al. 2020; Bakken et al. 2017; Coelho et al. 2017; Mekonnen and Hoekstra 2012). It is important to mention that the reservoirs of these HPPs have a hydrological function of storing water in the rain period, which supports the new run-of-river HPPs downstream in the dry period, guaranteeing operation throughout the year (Bueno et al. 2020).

3.3 Methodological Aspects for WFP Calculation in HPPs

The wide range of WFP estimates for the HPPs is explained by the local particularities, the design strategy of each facility, and, mainly, by methodological inconsistencies. Thus, it is possible to suggest the following guidelines for WFP examination in HPPs:

- Scientific consensus should be reached on using gross or net evaporation to calculate the WFP.
- The effects of large reservoirs that regularize flows should be considered, principally the benefit for hydropower generation in downstream plants, which can be designed as a run-of-river. This aspect can also be considered when WFP is calculated.

Methodologies that consider the reservoirs' multiple uses should be tested so that evaporation losses are apportioned among users. For instance, Berger and Finkbeiner (2010) pointed out that more detailed methodologies for WFP estimates of HPPs require high-resolution data in time and space. The method to be adopted should consider the study's objectives, accuracy, and purpose.

The temporal resolution of the datasets can significantly affect WFP estimates of HPPs. Annual estimates dampen the monthly variations in energy production and evaporation and, consequently, in the WFP estimation. The volume of water consumed when calculating the WFP of the HPPs is estimated based on evaporation from the reservoirs. This means that the water used for the construction and operation of the HPPs is disregarded, as well as for the extraction and production of materials involved in construction, for cooling equipment, washing machinery, and for use by employees, assuming that these uses are insignificant concerning the volume evaporated from the reservoir (Coelho et al. 2017; Hogeboom et al. 2018; Mekonnen et al. 2015).

The net WFP methodology appears more logical, as it enables us to evaluate the impact of altered land cover and land use on water availability in the watershed resulting from hydropower plant construction. Reservoir creation replaces vegetation with a water surface, thereby substituting evapotranspiration with reservoir evaporation (Bakken et al. 2015; Strachan et al. 2016). Nevertheless, considering net evaporation implies a conceptual error in the WFP calculation because, although the analysis is carried out in the same place, it is not at the same moment. In addition, the previous losses due to evapotranspiration would be associated with other uses (agriculture, for example) or native vegetation. The objective of the WFP is to measure water consumption for a specific purpose. Thus, it will no longer be available in time and space for another purpose.

Both approaches have their merits, depending upon the intended application of WFP estimates. Net WFP is suitable for analyzing changes in hydrology on a watershed scale, while gross WFP is recommended for assessments whose aim is to show the appropriated water volume for a given purpose and, therefore, it is unavailable for another purpose. The methodology for calculating WFP should take into account the multiple uses of the reservoirs, considering quantitative criteria (flows or withdrawal volumes) or the economic value of water (Bakken et al. 2016; Grubert 2016; Hogeboom et al. 2018; Scherer and Pfister 2016).

Nevertheless, we have neither a scientific definition nor a consensual acceptance of a methodology for allocating water losses due to evaporation from multiple uses of the reservoirs. Evaporation losses should be weighted among the different consumptive uses and not just linked to energy production since the reservoir's existence increases water availability in the surrounding flooded area. Also, the benefit to other users should be considered when estimating the WFP. In the case of multiple-use reservoirs, WFP calculation should consider weighing the economic benefits from non-consumptive water uses and not only consumptive uses (Zhao and Liu 2015). In the Zhao and Liu (2015) study, the economic value for the multiple uses of the Three Gorges hydropower plant reservoir in China was estimated based on market values (for power generation, water abstraction, navigation, and aquaculture), and damage costs, i.e., flood control due to the dam, were not considered.

Four methods for allocating water resources in multiple-use reservoirs were depicted in another study. It was concluded that allocation based on the proportion between the volumes of water withdrawn for each use was the most viable and robust approach. Economic allocation proved to be sensitive to the method used for economic water valuation. Including non-consumptive uses (e.g., flood control, recreation, navigation, fishing, and fish farming) in the WFP estimates are associated with a robust and reliable method (Bakken et al. 2016).

3.4 Water Footprint Estimates

Table 4 lists publications on the WFP of HPPs, covering plants worldwide. Most studies adopt the gross WFP methodology. The net WFP estimates are, on average, 30% of the gross WFP for those cases in which both methodologies were applied. The wide variation was mainly determined by the ratio between the reservoir area

Table 4 WFP in HPPs worldwide

Source	Gross WFP ($m^3.GJ^{-1}$)	Net WFP ($m^3.GJ^{-1}$)	Location
Torcellini et al. (2004)	18.9		120 plants in US
Mekonnen and Hoekstra (2011)	84.0		Eight plants in Brazil
Gerbens-Leenes et al. (2009)	22.3		Global average
Herath et al. (2011)	6.0	2.7	17 plants in New Zealand
Bueno et al. (2016)	130.0		1 plant in Southeast Brazil
Pasqualetti and Kelley (2008)	31.6		2 plants in Arizona, US
Mekonnen and Hoekstra (2012)	68.0		35 plants in different countries
Tefferi (2012)[a]	27.5		Ethiopia
Yesuf (2012)[a]	16.1	5.0	3 plants—Ethiopia
Liu et al. (2015)	3.6		209 plants China
Zhao and Liu (2015)	1.5		Three Gorges—China
Bakken et al. (2016)	10.5	0.05	2 plants—Norway
Scherer and Pfister (2016)	33.5	17.1	1473 plants in different countries
Strachan et al. (2016)	12.8	3.9	1 plant (Eastmain-1)—Canada
Coelho et al. (2017)	72.0	37.0	2 plants—North Brazil
Grubert (2016)	11.0	1.7	2200 plants—US
Hogeboom et al. (2018)	14.6		Global average (2235 reservoirs)
Adegun et al. (2018)	381.3		1 plant—Nigeria

[a]Cited by Bakken et al. (2013). Cited in Bueno (2018)

and installed capacity of the plants. Plants with the highest ha.MW^{-1} ratio showed the largest WFP (Bueno et al. 2016). WFP was estimated for the Camargos HPP upstream of the Funil HPP based on power generation data provided by the plant operator and meteorological data monitored close to the shore of the reservoir from 2010 to 2014. The average gross WFP for the studied period was equivalent to 130 $m^3.GJ^{-1}$. It was found that during the critical period of drought observed in Southeast Brazil (2013/2014) (Silva and Mello 2021), the WFP of the Camargos HPP was very high, reaching 272 $m^3.GJ^{-1}$ in 2014 due to low energy production. As in this case, we have a cascade of HPPs, WFP can also be considered both facilities, being Camargos fundamental for accumulation water that will support the run-of-river aspect of the Funil.

4 Study Case: WFP of the Funil HPP, Southeast Brazil

4.1 General Aspects

The Funil HPP (official name: Eng. José Mendes Júnior), with an installed power of 180 MW, is located on the Grande River, at the upper basin, in the southern Minas Gerais state. The Grande River basin is important for hydroelectric generation in

Brazil, with a capacity of 7619 MW (Bueno et al. 2020). The following section will use the Funil HPP to illustrate the analysis and computations necessary to estimate WFP using meteorological data across the reservoir and 2 years of operational data (July 2015 to June 2017). In summary, the objectives are:

- Estimate the Funil HPP reservoir's daily, monthly, and annual evaporation rates from data monitored at meteorological stations installed on the reservoir's shore.
- Quantify this HPP's WFP based on evaporation estimates from the reservoir and electricity generation provided by the plant's operator, comparing our results with other HPPs worldwide.
- Analyze the intra-annual behavior of the Funil HPP's WFP.
- Evaluate the sustainability of the Funil HPP's WFP for electricity production, considering the local and regional scales.

4.2 Methodology

4.2.1 Geographical Location

The Funil HPP has a dam that is 50 m high (Fig. 1), located on the Grande River in the southern Minas Gerais state, Brazil, at the geographical coordinates of 21°08′35″ S and 45°02′10″W. The plant is located in the upper Grande River basin, with a contributing area to the reservoir of 15,716 km^2. It is the third hydropower project on this river (from headwater in the Mantiqueira Range). According to the National Water Resources Plan (BRASIL 2006), the Grande River basin embraces a total area of 143,173 km^2, comprising parts of the Minas Gerais and São Paulo states, being one of the main rivers that form the Paraná Hydrographic Region (Fig. 1).

The Funil HPP is a run-of-river plant (constant-level), meaning that its reservoir does not have the function of regularizing inflows. It has a useful volume of 6.8 hm^3, with an operating water level between 807.8 and 808.0 m asl, an average depth of

Fig. 1 Geographical location of the Funil HPP in the Grande River basin and the Paraná Hydrographic Region and a view of the Funil Hydropower Plant facility

7.7 m, and a quasi-constant water surface of 34.5 km^2. The climate in the region is temperate (Cwa), with humid summers and mild temperatures and dry, cold winters (Mello et al. 2021). The average annual rainfall is around 1450 mm, with 85% of the rainfall occurring between November and March. The average annual temperature is around 20 °C, with average minimum and maximum temperatures ranging from 9 to 24 °C, respectively (Mello et al. 2021).

4.2.2 Electric Energy Production at Funil HPP

The Funil HPP's average power generation from July 2015 to June 2017 was 63 MW, ranging from 23 to 174 MW. The generation pattern over the studied period is presented in Fig. 2 (a. daily; b. monthly), with the lowest values in July and August and the highest generation seen in December and January.

4.2.3 Affluent Streamflow to the Funil HPP

The main tributaries flowing into the Funil HPP reservoir are the Grande River (monitored at the Itutinga gauging station), the Mortes River (Ibituruna gauging station), and the Capivari River (Itumirim gauging station). It should be noted that the Camargos HPP reservoir regulates the flows of the Grande River, while the flows of the Mortes River and Capivari River are not influenced by reservoirs (Fig. 3).

The monitoring stations control 90% of the drainage area into Funil reservoir. Thus, the inflow was estimated using a factor defined by the ratio between the total contributing area and the sum of the sub-basins delineated at the stations. In Fig. 3, we can also see the daily streamflow and the monthly inflow averages to the Funil HPP reservoir, with an average of 193 m^3.s^{-1} for the studied period. The most significant contribution was from the Grande River (49%), the Mortes River (40%), and the Capivari River (11%).

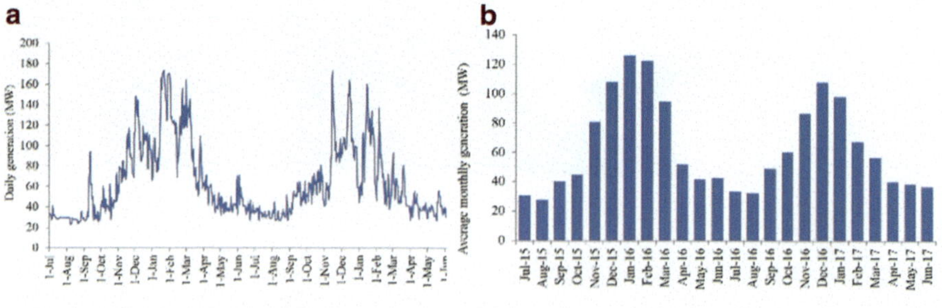

Fig. 2 Daily (**a**) and monthly (**b**) average power generation in the Funil HPP between July 2015 and June 2017

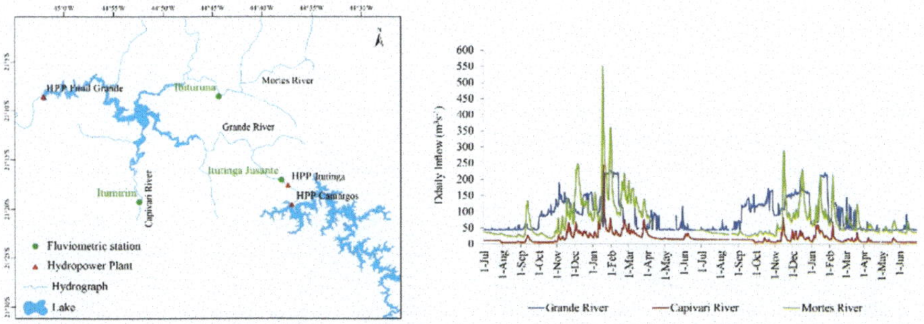

Fig. 3 Rivers, fluviometric gauging stations that monitor the affluent streamflow to the Funil HPP reservoir, and the daily inflows to Funil HPP in the studied period

4.2.4 Meteorological Datasets Monitored Throughout the Studied Period

Weather monitoring was carried out near the reservoir over a period of 2 years, from July 2015 to June 2017, using four automatic weather stations (Fig. 4) to support the estimation of evaporation and evapotranspiration fluxes from Funil reservoir and the flooded areas. Air temperature and relative humidity data, wind speed and direction, incident solar radiation, precipitation, and atmospheric pressure were monitored hourly. The daily average values represented the atmospheric boundary layer over the surface reservoir, which is reasonable considering that the stations were installed very close to the reservoir (Fig. 6). These averages were used in the evaporation and evapotranspiration estimates.

4.2.5 Temperature Monitoring of the Funil Reservoir

In addition to the meteorological stations, three sensors (Fig. 5a) were installed at different locations in the Funil reservoir to monitor the surface water temperature at hourly intervals in 2016. From April to June 2017, the vertical profile of the reservoir's temperature was monitored only at the point corresponding to the "Tw 3" sensor (Fig. 5a). Based on the preliminary results obtained during the initial phase of this monitoring, the water sensors were gathered at just one location so that we could have temperature monitored in deeper layers, as shown in Fig. 5b.

4.3 Evaporation Calculation for the Funil HPP Reservoir

The purpose of this section is to present the Penman–Monteith method, improved by McJannet et al. (2008), which was used in this study for estimating evaporation from

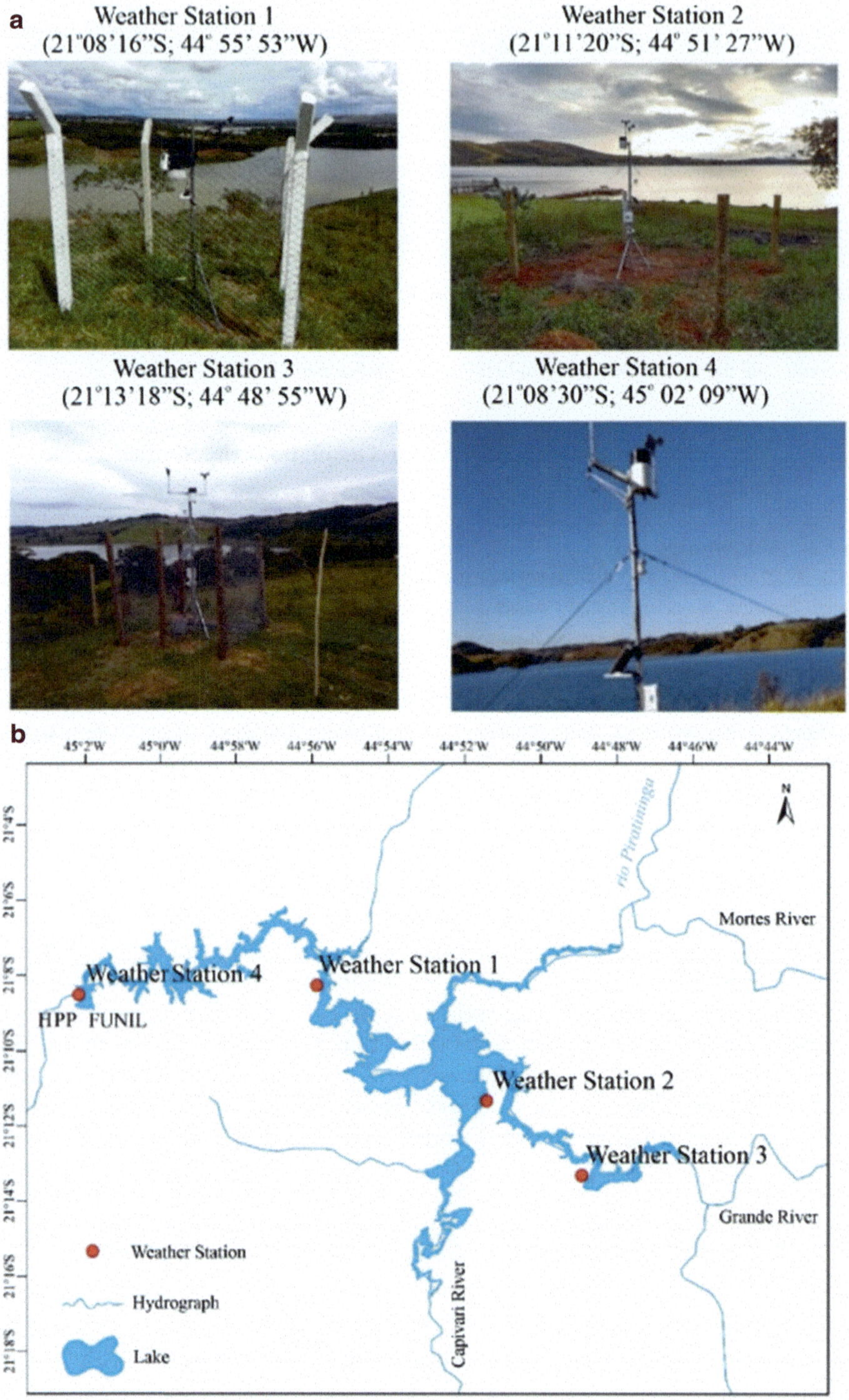

Fig. 4 Weather stations installed near the Funil HPP reservoir (**a**. view and **b**. location throughout the reservoir). (Source: Adapted from Bueno 2018)

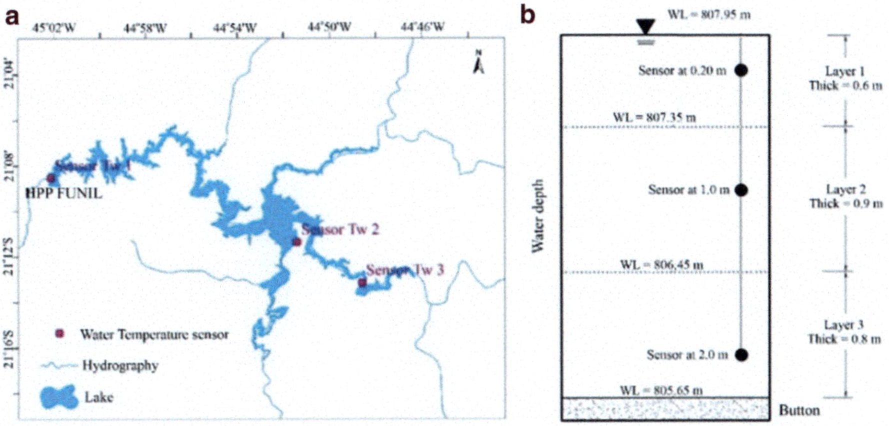

Fig. 5 Location of the water sensors for temperature measurement in the Funil HPP reservoir (**a**) and scheme designed for measurement in deeper layers (**b**)

reservoirs. This is relevant as these equations are not found in traditional hydrology books, and the readers can easily access them in this chapter. The basic equation is:

$$E = \left(\frac{\Delta_{\mathrm{w}}(R_{\mathrm{L}} - G) + \gamma f(u)(ew - ea)}{\Delta_{\mathrm{w}} + \gamma} \right) \frac{1}{\lambda}, \tag{1}$$

where E is the evaporation (mm.day^{-1}); λ is the water vaporization latent heat at air temperature (MJ.kg^{-1}); Δw is the slope of the saturation water vapor pressure at water temperature (kPa.°C^{-1}); R_{L} is the radiation balance (MJ.m^{-2}.day^{-1}); G is the heat storage variation in the reservoir (MJ.m^{-2}.day^{-1}); $f(u)$ is a function associated with wind velocity and reservoir's area (MJ.m^{-2}.day^{-1}.kPa^{-1}); ew is the water vapor saturation pressure at water temperature (kPa); ea is water vapor partial pressure at air temperature (kPa); and γ is psychrometric constant (kPa.°C^{-1}).

To calculate the evaporation using Eq. 1, the following data are needed: area and average depth of the reservoir; air temperature and relative humidity; wind speed at a 2-m height; incident solar radiation; and atmospheric pressure. The calculation sequence was used to estimate evaporation from the Funil HPP.

$$\lambda = 2.501 - \left(2.361 \times 10^{-3}\right) T_{\mathrm{a}}, \tag{2}$$

where λ is the latent heat of vaporization (MJ.kg^{-1}); T_{a} is the average air temperature (°C). For the normal air temperature and pressure conditions, λ is approximately 2.45 MJ.kg^{-1}.

$$\gamma = \left(1.63 \times 10^{-3}\right)\frac{\mathrm{P}}{\lambda} \tag{3}$$

Here, γ is the psychrometric constant (kPa.°C^{-1}); P is the atmospheric pressure (KPa).

$$es = 0.6108\mathrm{e}^{\left(\frac{17.27T_{\mathrm{a}}}{T_{\mathrm{a}}+237.3}\right)} \tag{4}$$

Here, es is the water vapor saturation pressure at air temperature (kPa).

$$ea = \frac{es\mathrm{UR}}{100} \tag{5}$$

Here, ea is the partial water vapor pressure at air temperature (kPa); UR is the relative air moisture (%).

$$ew = 0.6108\mathrm{e}^{\left(\frac{17.27T_{\mathrm{w}}}{T_{\mathrm{w}}+237.3}\right)} \tag{6}$$

Here, ew is the water vapor saturation pressure at water temperature (kPa); T_{w} is the water temperature (°C).

$$\Delta\mathrm{w} = \frac{4098es}{\left(T_{\mathrm{w}} + 237.3\right)^2} \tag{7}$$

Here Δw is the slope of the air vapor saturation pressure curve at water temperature (kPa.°C^{-1}).

$$f(u) = (3.80 + 1.57u_{10})\left(\frac{5}{\mathrm{A}}\right)^{0.05} \tag{8}$$

Here, $f(u)$ is the aerodynamic function (MJ.m^2.day^{-1}.kPa^{-1}); A is the reservoir area (km^2); u_{10} is the wind velocity at a 10-m height (m.s^{-1}), estimated from wind velocity at a 2-m height (u_2):

$$u_{10} = \frac{1}{0.748}u_2 \tag{9}$$

The radiation balance "R_{L}" is the sum of the short-wave radiation balance and longwave radiation balance, calculated as follows (McJannet et al. 2008).

$$R_{\mathrm{L}} = R_{\mathrm{s}}(1 - a) + (R_{\mathrm{i}} - R_{\mathrm{o}}) \tag{10}$$

R_{L} is the radiation balance (MJ.m^{-2}.day^{-1}); R_{s} is the incident solar radiation (MJ.m^{-2}.day^{-1}); a is the reservoir surface albedo (assumed as 0.08); R_i is input longwave radiation (MJ.m^{-2}.day^{-1}); R_o is the output longwave radiation (MJ.m^{-2}.day^{-1}).

$$R_i = 1 - (1 - C_f)\left(0.261 e^{\left(-7.77\text{x}10^{-4} {T_a}^2\right)}\right)\sigma(T_a + 273.15)^4 \tag{11}$$

Here, σ is the Stefan–Boltzmann constant (4.903 × 10^{-9} MJ.m^{-2}.dia^{-1}.K^{-4}); C_f is the cover fraction by clouds (adimensional), given by:

$$C_f = 1.1 - \left(\frac{R_s}{(0.75 + (2 \times 10^{-5} h) R_a}\right) \tag{12}$$

where h is the elevation (m); R_a is the radiation at the top of the atmosphere (MJ.m^{-2}.day^{-1}), calculated by:

$$R_a = 37.586 dr(\omega s \text{sen}\phi \text{sen}\delta + \cos\phi \cos\delta \text{sen}\omega s). \tag{13}$$

Here, φ is the latitude (radians, negative in the South Hemisphere); δ is the solar declination (radians); ωs is the hourly angle of the sunrise and sunset (radians); dr is the Earth–Sun relative distance, defined as:

$$\delta = 0.4093 \text{sen}\left(\frac{2\pi}{365} J - 1.405\right) \tag{14}$$

$$\omega s = \arccos(-\tan\phi \tan\delta) \tag{15}$$

$$\text{dr} = 1 + 0.033 \cos\left(\frac{2\pi}{365} J\right). \tag{16}$$

J is the Julian day (adimensional). The part of the longwave radiation out (R_o) in the Penman–Monteith energy balance is estimated by (McJannet et al. 2008):

$$R_o = \varepsilon_w \sigma (T_w + 273.15)^4 \tag{17}$$

where ε_w is the water emissivity (0.97); T_w is the water surface temperature (°C), estimated by:

$$T_{w,i} = T_e + (T_{w,i-1} - T_e) e^{\left(-\frac{1}{\tau}\right)} \tag{18}$$

$T_{w,i}$ and $T_{w,\ i\ -\ 1}$ are the water temperature on the day i and $i - 1$ (°C); τ is the time constant (days); T_e is the equilibrium temperature (°C) (De Bruin 1982), calculated as follows:

$$T_e = T_n + \frac{R_L^*}{4\sigma(T_n + 273.15)^3 + f(u)(\Delta_n + \gamma)} \tag{19}$$

where $f(u)$ is the aerodynamic function (MJ.m^{-2}.day^{-1}.kPa^{-1}). T_n is the wet bulb (°C); Δ_n is the slope curve of the saturation pressure at wet bulb temperature (kPa. °C^{-1}); R_L^* is the balance radiation at wet bulb temperature (MJ.m^{-2}.day^{-1}).

$$T_n = \frac{0.066Ta + Td\left(\frac{4098e_a}{(Td+237.3)^2}\right)}{0.066 + \left(\frac{4098e_a}{(Td+237.3)^2}\right)} \tag{20}$$

T_d is the dew temperature (°C), estimated by:

$$T_d = \frac{237.3\text{Log}\left(\frac{ea}{0.6108}\right)}{7.5 - \text{Log}\left(\frac{ea}{0.6108}\right)} \tag{21}$$

$$\Delta_n = \frac{4098\left(0.6108e^{\left(\frac{17.27Tn}{Tn+237.3}\right)}\right)}{(Tn + 237.3)^2} \tag{22}$$

$$R_L^* = Rs(1 - a) + (R_i - R_{on}). \tag{23}$$

R_s is the incident solar radiation (MJ.m^{-2}.day^{-1}), measured at the meteorological station; R_{on} is the longwave out radiation (MJ.m^{-2}.day^{-1}) at the wet bulb temperature:

$$R_{on} = \sigma(T_a + 273{,}15)^4 + 4\sigma(Ta + 273{,}15)^3(T_n - T_a) \tag{24}$$

The time constant (τ, days) is estimated using (De Bruin 1982):

$$\tau = \frac{\rho_w c_w p}{4\sigma(T_n + 273.15)^3 + f(u)(\Delta_n + \gamma)} \tag{25}$$

where c_w is the water-specific heat (0.0042 MJ.kg^{-1}.°C^{-1}); p is the reservoir's average depth (m).

The variation of the reservoir heat storage (G in MJ.m^{-2}.day^{-1}), needed in the energy balance of the Penman–Monteith method (Eq. 2), is estimated by (Finch 2001):

$$G = \rho_w c_w p(T_{w,i} - T_{w,i-1}) \tag{26}$$

where ρ_w is the water density (1000 kg.m^{-3}); c_w is the water-specific heat (0.0042 MJ.kg^{-1}.°C^{-1}).

4.4 Storage Heat Variation in the Funil HPP Reservoir

The variation in heat stored in the reservoirs (G) is obtained using Eq. 26 as a function of the water temperature. The most accurate way to estimate G is using measurements of the vertical profile of the water temperature (Giadrossich et al. 2015). These daily interval calculations were carried out using Eq. 27 (Giadrossich et al. 2015).

$$G(t) = \frac{\rho_w c_w}{A(t)} \sum_{i=1}^{n_i} V_i \Delta T_i \tag{27}$$

where $G(t)$ is the heat variation in the reservoir (MJ.m^{-2}.day^{-1}) on the day t; $A(t)$ is the reservoir's area in function on day t (km^2); n_i is the number of layers monitored (three in this study case); ΔT_i is the water temperature variation observed at depth i, between the days t and $t - 1$; V_i is the reservoir's volume corresponding to depth i, dependent on layer thickness (hm^3). $A(t)$ and V_i are taken from the volume–area–depth curve of the reservoir's curve.

- $A(t)$ = obtained using the volume–area–depth curve and the reservoir depth measurement, which is recorded daily. As FPP is a runoff-river reservoir, $A(t)$ is almost constant over time (34.5 km^2).
- $\Delta T_i = T(t) - T(t - 1)$, calculated based on the water temperature taken at different depths.
- V_i = Table 5.

The evapotranspiration estimates in the area flooded by the Funil HPP reservoir were carried out using the Penman–Monteith method, taking into account the characteristics of each land use and land cover class before the filling of the reservoir (Allen et al. 1998):

$$\text{ETP} = \left[\frac{\Delta(R_L - G) + \rho_a.cp\frac{(es - ea)}{ra}}{\Delta + \gamma\left(1 + \frac{rs}{ra}\right)}\right]\left(\frac{fc}{\lambda.\rho_w}\right) \tag{28}$$

ETP is the potential evapotranspiration (mm.day^{-1}); Δ is the water vapor pressure curve at air temperature (kPa.°C^{-1}); es is the water vapor saturation pressure (kPa); ea is the partial water vapor pressure (kPa); R_L is the net radiation (MJ.m^{-2}.s^{-1}); ra

Table 5 Volumes of the layers corresponding to the points of the water temperature monitoring in the Funil HPP reservoir

Layer	Superior level (m)	Inferior level (m)	Superior volume (hm^3)	Inferior volume (hm^3)	Volume (hm^3)
1	807.950	807.350	267.23	246.81	20.42
2	807.350	806.450	246.81	218.86	27.95
3	806.450	805.650	218.86	196.37	22.49

is the aerodynamic resistance (s.m^{-1}); *rs* is the surface resistance (s.m^{-1}); *G* is the heat flux from the ground; ρ_a is the air density (1.08 kg.m^{-3}); ρ_w is the water density (1000 kg.m^{-3}); λ is the latent heat of vaporization (2.45 MJ.kg^{-1}); *cp* is specific heat of the moist air (0.001013 MJ.kg^{-1}.°C^{-1}); γ is the psychrometric constant (0.0613 kPa. °C^{-1}); *fc* is the unit factor conversion (m.s^{-1} to mm.day^{-1}) (8.64.10^{7}). In order to estimate the actual evapotranspiration, it is necessary to know the area and the crop factor (*Kc*) of each land use class before the flooding.

4.5 Gross and Net WFP Calculation for the Funil HPP Reservoir

The main difference between these methodologies lies in the water loss considered as consumption for energy generation. The gross WFP method consists of reservoir evaporation divided by energy production on an annual or monthly scale, given by the following equation:

$$\text{Gross WFP} = \frac{E}{H} \tag{29}$$

where WFP is the water footprint (m^3.GJ^{-1}); *H* is the annual electric energy production (GJ.year^{-1}); *E* is the Funil's reservoir evaporation (m^3.year^{-1} or per month), calculated by multiplying depth evaporation (m) and reservoir area (m^2).

The second approach considers water losses due to evapotranspiration from the vegetation cover in the area flooded by the reservoir, i.e., considering the concept of net evaporation (ONS 2004). Thus, the net WFP method reflects an incremental impact in analysis with and without a reservoir, i.e.:

$$\text{Net WFP} = \frac{(E - \text{ETR})}{H} \tag{30}$$

ETa is the actual evapotranspiration (m^3.year^{-1} or monthly), obtained by multiplying the depth of actual evapotranspiration (m) and the reservoir area (m^2).

5 WFP in the Funil HPP

In this section, we present the results obtained for the study conducted with the Funil HPP following the methodological procedure presented in this chapter. It is presented from the meteorological characterization in the study period (database for WFP calculation) to the WFP temporal behavior

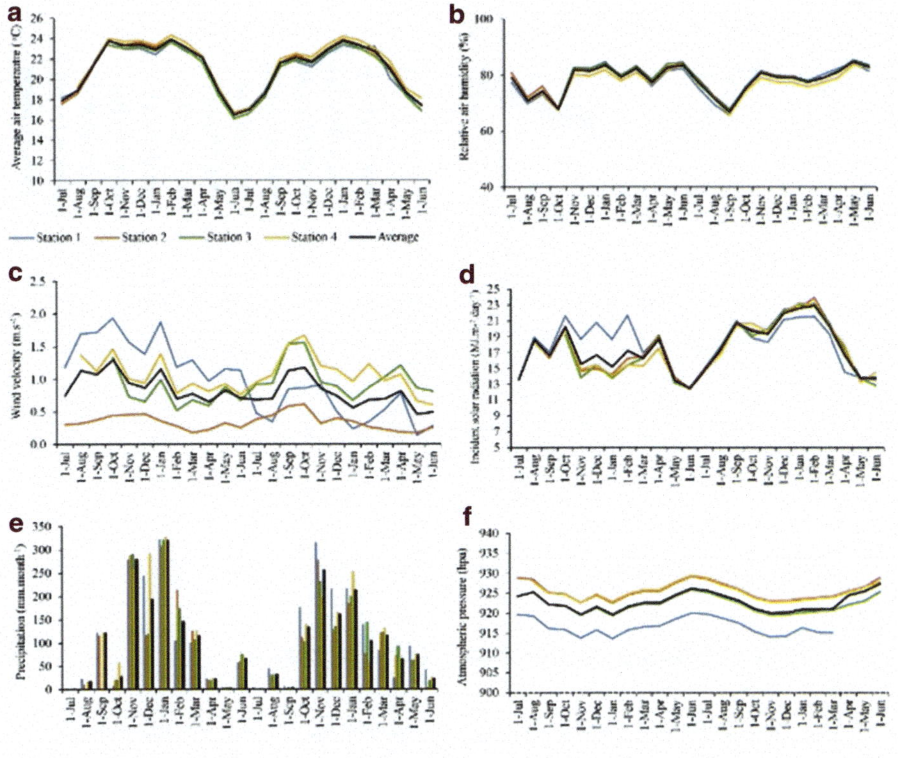

Fig. 6 Meteorological behavior in the Funil HPP reservoir between July 2015 and June 2017

5.1 Meteorological Characterization of the Funil HPP Reservoir

The behavior of the weather between July 2015 and June 2017 is in Fig. 6, which shows the monthly average air temperature, relative humidity, wind speed, incident solar radiation, monthly amount of precipitation, and atmospheric pressure. In addition, the averages of these data observed at the four weather stations installed around the Funil HPP reservoir are also in this figure (see Fig. 5 for the location of each station). Some aspects are relevant, like the well-characterized dry and wet periods in the region (Fig. 6e), which is very important for energy production planning, the high variability of the wind, implicating different behavior of the evaporation (despite the reservoir being small if compared with the other facilities in Brazil, it is relevant we have more samples of the weather along the reservoir), and similarities between the air temperature and humidity between the stations. All these weather datasets were used in the reservoir evaporation calculation and influenced the WFP of the Funil HPP.

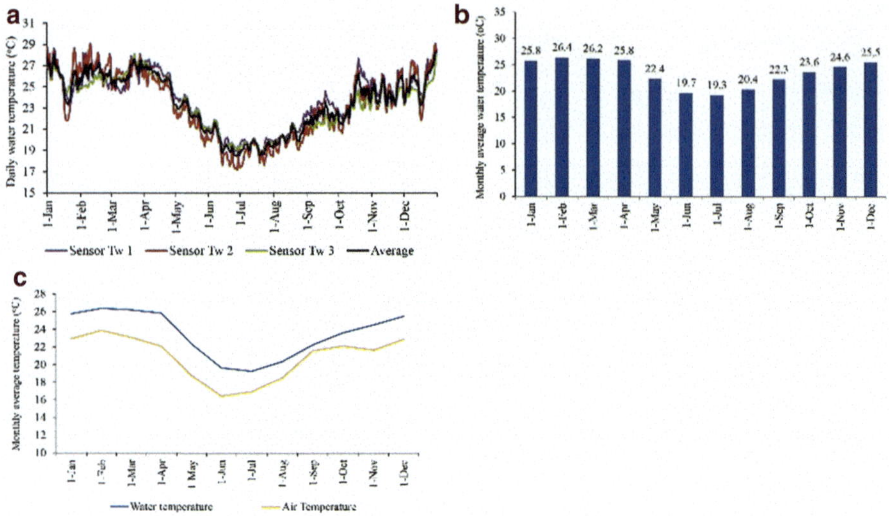

Fig. 7 Daily (**a**) and monthly (**b**) water temperature in the Funil HPP reservoir in 2016 and comparison of the water temperature (**c**)

5.2 Thermodynamic of the Funil HPP Reservoir

Following the methodological procedure previously described, the thermodynamic of the reservoir has a fundamental relevance for the reservoir's evaporation because the water temperature is indispensable in this estimation. Thus, Fig. 7a illustrates the daily surface water temperature observed at the three sensors installed in the Funil HPP reservoir throughout 2016 (see schema in Fig. 5). The temperature behavior at these three locations in the reservoir was very close, with average daily and monthly differences less than 1.0 °C and 0.5 °C, respectively. Figure 7b shows the monthly averages of the surface water temperature of the Funil HPP reservoir, obtained by averaging the readings from the three sensors, indicating the lowest temperatures in June and July and the highest in February and March, approximately following the air temperature but with a delay in the summer as January and February are the hottest months

The surface water temperature was higher than the air temperature, with an average difference of 2.5 °C throughout 2016, representing a condition of sensible heat flux to the atmosphere (positive). Surface water temperature showed the same seasonal distribution pattern as air temperature in Fig. 7c, with minimum values in winter and maximum values in summer, but as commented before, with a slight delay between them. Due to its greater thermal capacity, the water temperature shows much more attenuated oscillations than the air temperature. Another relevant aspect is that we can observe a significant difference between the water temperature in the summer and in the winter, reaching 5 °C in the mean average value. This is relevant because in the tropical/subtropical regions, water temperature has been considered insignificant for the calculation of evaporation, where an average value has been taken instead of the observed one.

5.3 Heat Storage Variation (G) in the Funil HPP Reservoir

Water temperature also influences the *G* (variation in heat stored in the reservoir), a portion of the energy balance used to calculate evaporation. The average (daily and monthly) *G* values calculated from the vertical water temperature profile are shown in Table 6. There is a considerable difference in these three consecutive months, mainly with a strong reduction in May.

Negative *G* values mean that, at this time of year, heat is being released from the reservoir (lower water temperature), contributing to the energy balance, i.e., more energy is available for evaporation. The daily behavior of *G* based on the reservoir water temperature is in Fig. 8, where we can a large wide values in consecutive days. These datasets demonstrate how relevant it is to consider the water temperature for evaporation calculation, which cannot be considered irrelevant in most of the WFP studies in HPPs.

Table 6 Examples of heat storage variation in the Funil HPP reservoir (*G* portion of the balance energy) calculated using the observed water temperature

	Heat storage variation (*G*)	
Date	$MJ.m^{-2}.day^{-1}$	$MJ.m^{-2}.month^{-1}$
April 2017	−0.8	−20.9
May 2017	−0.3	−9.6
June 2017	−0.6	−16.8

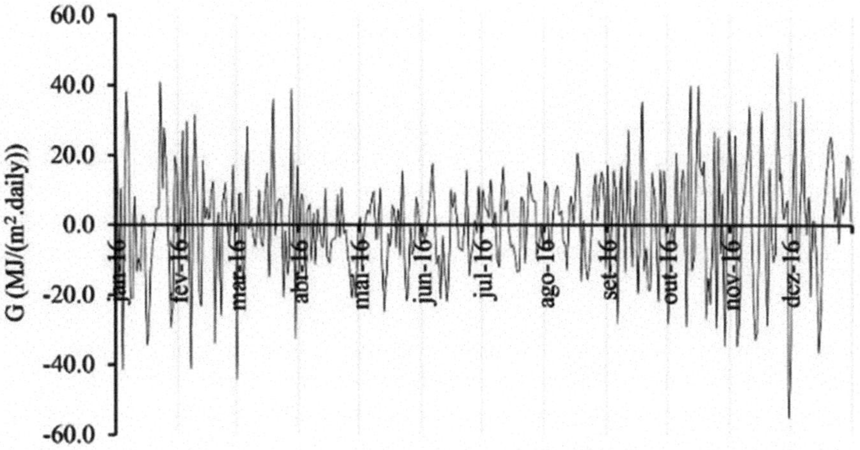

Fig. 8 *G* behavior throughout 2016 based on the observed water temperature in the three sensors

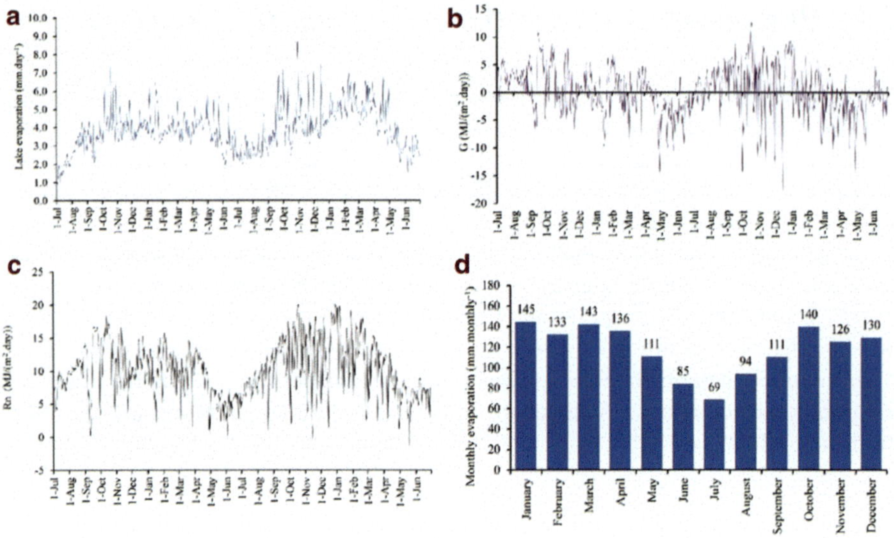

Fig. 9 Daily evaporation (**a**), *G* component (**b**), net radiation (**c**) monthly, and (**d**) evaporation between July 2015 and June 2017

5.4 Funil HPP Reservoir Evaporation

The average daily evaporation of the Funil HPP reservoir from July 2015 to June 2017 was 3.9 mm.day^{-1}, ranging from 0.9 to 7.5 mm.day^{-1}. Figure 9 shows the temporal behavior of daily evaporation from the Funil reservoir (a), *G* temporal behavior (b), and net radiation (c). The variation on a daily scale is associated with different meteorological conditions throughout the study period. This means that the maximum and minimum evaporation values are very sensitive to field measurements of extreme air temperature and humidity, wind speed, and solar radiation, which depend on local and momentary conditions, such as rainfall and cloudy or clear sky. Figure 11d presents the average monthly evaporation of the Funil HPP reservoir during the studied period, and the mean annual value was 1422 mm.year^{-1}.

Evaporation from the Funil HPP reservoir follows the temporal distribution pattern of the variables that influence the process, especially *G* and solar radiation. Thus, in the months with the highest incident solar radiation and air temperature (December to February and October), the highest evaporation rates were obtained, while in the months with the lowest radiation (June and July), the lowest evaporation values were obtained. Wind speed and relative humidity also influence evaporation but with less sensitivity in the Penman–Monteith method. For the Funil HPP reservoir, approximately 80% of the estimated evaporation comes from the energy balance, especially the radiation balance. The mass transfer term (aerodynamic) weighs less in the evaporation estimate.

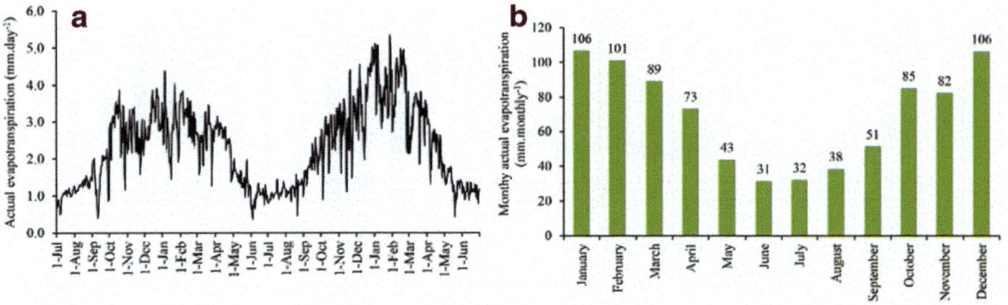

Fig. 10 Daily (**a**) and monthly (**b**) actual evapotranspiration of the area flooded by the Funil HPP reservoir

5.5 *Actual Evapotranspiration of the Area Flooded by the Funil HPP Reservoir*

Using the Penman–Monteith method, the actual evapotranspiration of the area flooded by the Funil HPP reservoir showed an average daily value of 2.3 mm. day^{-1}, varying from 0.4 to 5.4 mm.day^{-1}. Figure 10a illustrates the temporal distribution of the flooded area's daily evapotranspiration during the study period. Figure 10b shows the monthly average actual evapotranspiration, with a temporal distribution similar to the seasonal pattern of incident solar radiation, with minimum rates in June and July and maximum rates in December and January. Based on these values, the actual evapotranspiration of the area flooded by the Funil HPP reservoir was estimated at 837 mm.year^{-1}, equivalent to 67% of the average annual rainfall observed in the studied period.

5.6 *Net Evaporation from the Funil HPP Reservoir*

Net evaporation is calculated using the monthly values of E and ETa, and it can be estimated for the Funil HPP reservoir by subtracting ETa from E. The average daily value is 1.6 mm.day^{-1}, ranging from −0.7 to 5.6 mm.day^{-1}. Figure 11 shows the behavior of the reservoir's net evaporation on the daily (a) and monthly (b) scales and a comparison between evaporation and actual evapotranspiration (c).

We can observe some days, mainly January 2017, with negative net evaporation values, i.e., actual evapotranspiration (ETa) greater than evaporation (E). Since the meteorological data used to calculate E and ETR are the same, the negative net evaporation values refer to the variation in the reservoir heat stored (G), which is linked to the water temperature. When calculating evaporation, the adapted Penman–Monteith method considers the G participation in the energy balance. In the case of actual evapotranspiration, the Penman–Monteith method disregards it, assuming that the energy flux to the soil on a daily scale is zero (Allen et al. 1998). Therefore, the days of negative net evaporation correspond to those with the highest (positive)

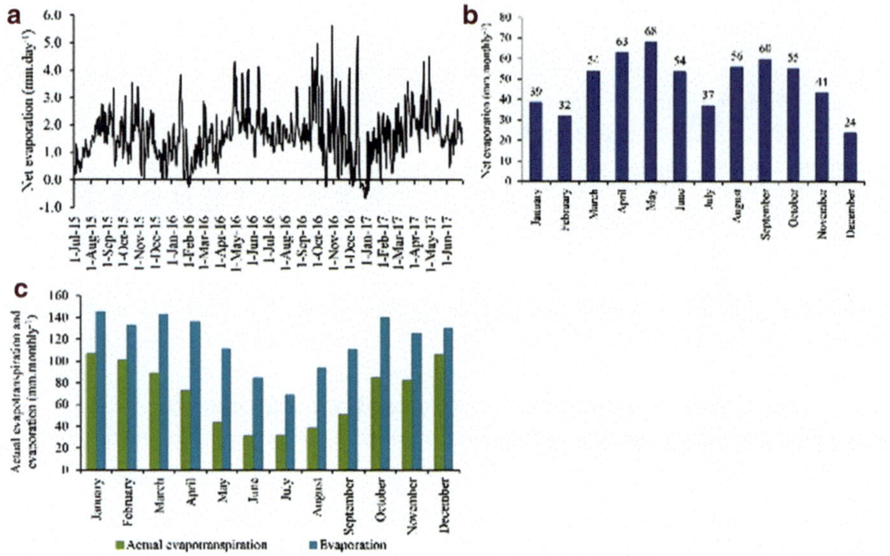

Fig. 11 Daily (**a**) and monthly (**b**) net evaporation from the Funil HPP reservoir and the average evaporation and actual evapotranspiration (**c**)

values of *G*, i.e., when the reservoir stores more energy from the incident solar radiation, reducing the energy available for the evaporation process.

Figure 11c compares the average monthly evaporation and actual evapotranspiration for the studied period. A similar temporal distribution of these variables can be seen, which is related to the seasonal behavior of solar radiation. On a monthly scale, the net evaporation values are positive, i.e., the reservoir's evaporation exceeds the actual evapotranspiration from the flooded area. The highest net evaporation was calculated in May due to a combination of high evaporation and low evapotranspiration. The lowest net evaporation was obtained in December because the evapotranspiration for this month is also high and close to the reservoir's evaporation.

5.7 *Water Footprint at the Funil HPP Reservoir*

The monthly estimates and the intra-annual variation of the gross and net WFPs of the Funil HPP in the study period are presented in Fig. 12. There is a similar temporal distribution between the gross and net WFPs and an inverse behavior concerning energy production.

The average monthly WFP of the Funil HPP was 29 and 13 $m^3.GJ^{-1}$, respectively, for gross and net WFPs. Net WFP corresponds, on average, to 45% of gross WFP when considering evapotranspiration in the area flooded by the Funil reservoir. This difference aligns with other studies considering both WFPs (Bakken et al. 2013;

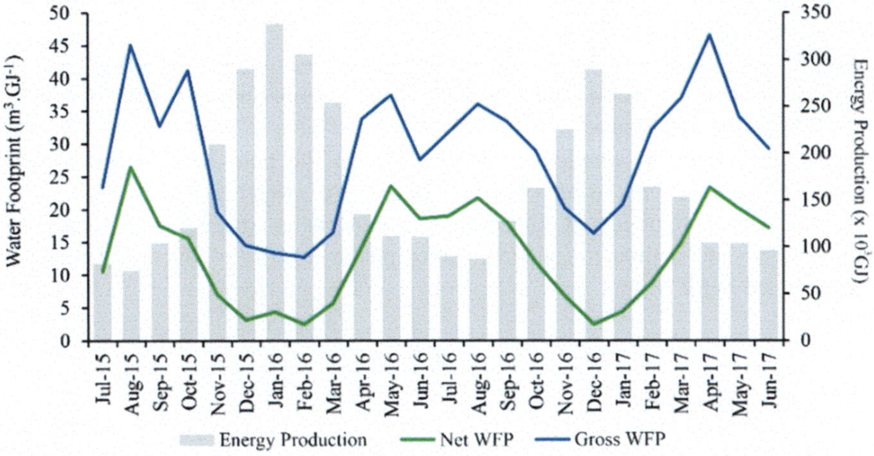

Fig. 12 Gross and net WFP of the Funil HPP between July 2015 and June 2017 and the respective electrical energy produced

Coelho et al. 2017); however, the accuracy of the WFP value is higher than such studies due to the methodology, based on the field observations.

The gross WFP of the Funil HPP is mainly related to energy generation, with decreasing values from August to December, when there is a progressive increase in energy production. Increased WFP values in the first semester of the year can be seen due to reduced energy production. It is important to highlight that summer is the season of the year in which the Brazilian electrical system is more required because in this season, there is a significant impact of the use of air conditioning, freezers, and other devices by the industry and agriculture in the country to keep the production in high quality. From November to March, the gross WFP is below average, while from April to October, it is above average. In June and July, despite lower evaporation, the gross WFP was close to the average due to low energy production. The gross WFP peak is seen in August and is a consequence of the lower energy production. In April, the gross WFP was also high and is associated with relatively low energy production and high evaporation from the Funil HPP reservoir. On the other hand, even with high evaporation rates due to increased solar radiation, Funil HPP had the lowest gross WFP in December and January because of the peak in energy production.

Net WFP has shown similar behavior as gross WFP, i.e., a decrease from August to December and an increase from January to May due to a reduction in energy production. From December to February, net WFP showed the lowest values (minimum in December) due to higher energy production and lower net evaporation. In May and August, net WFP reached its maximum values (with a peak in August) due to low energy production and higher net evaporation. From November to March, the net WFP was below average; from April to October, it was above average, just like the Gross WFP. Despite low net evaporation (lowest solar radiation) in July, net WFP was above average due to low energy production.

The WFP of the Funil HPP was lower than estimates obtained for other HPPs in Brazil and around the world (Bakken et al. 2013; Hogeboom et al. 2018; Liu et al. 2015; Mekonnen and Hoekstra 2012). However, we have to keep in mind that only two specific years were employed in the study, as the purpose is to test a more accurate procedure, i.e., using observed datasets of weather and actual energy production. Studies conducted considering several plants and regions typically use average values and estimated (or the installed power) energy production, leading to overestimated WFP.

6 Considerations

6.1 Main Factors Influencing WFP in the Funil HPP

The factors influencing the WFP of HPPs should combine weather and plant operational aspects. Evaporation correlates with the WFP, i.e., the more intense this process is, the greater the WFP is, being inverse with electricity production. The WFP of the Funil HPP was more dependent on energy production than evaporation since the latter showed less temporal variability than the former. The Funil HPP's power generation showed greater monthly oscillations than evaporation and net evaporation. It is important to highlight that Funil HPP's operation depends on the Camargos HPP operation (Fig. 4). This plant is upstream of Funil, and its reservoir is of the accumulation type, i.e., it has significant oscillation throughout the year, accumulating water in the rainy season and discharging in the dry season to support the Funil HPP reservoir operating like a run-of-river. In this way, the WFP showed monthly variability due to the fluctuations in energy production throughout the year, which is linked to the market demand for energy and the Camargos Reservoir operation.

Notably, the gross WFP is highly sensitive to evaporation for HPPs, with a more significant variation in the reservoir's area (Bueno et al. 2016). The Funil HPP operation is related to the Camargos HPP operation with decreasing gross and net WFP during the reservoir's depletion period (from August to December), when there is a reduction in the useful volume due to higher outflows. Then, an increasing WFP, associated with the filling of the Camargos Reservoir from January to April, was observed (Bueno et al. 2016) because of the increase in volume provoked by a reduction in outflows, creating an increased water surface and solar radiation. Therefore, the WFP should be assessed considering the cascade of plants, as there is hydrological and operation synergism between the reservoirs throughout the year (Bueno et al. 2020).

6.2 *WFP Sustainability at Funil HPP*

The assessment of the WFP sustainability (Hoekstra et al. 2011) needs to consider, among other aspects, the availability and demand datasets and the water balance of the area covered by the study. An initial analysis showed that the Upper Grande River basin, where the Funil HPP is located, is in a comfortable water demand situation. The water consumption in the basin and sub-basins has been lower than the water availability (e.g., greater the streamflow used for a grant), and no critical regions are identified for meeting the uses of water resources (Bueno et al. 2020). Thus, observing the sustainability analysis criteria (Hoekstra et al. 2011), the basin's water availability satisfactorily covers the demand for sustaining the ecological streamflow and consumptive uses.

Additionally, in a local analysis, the assessment can be made to compare the water volume evaporated from the reservoir and the precipitation on it. Thus, during the monitoring period (July 2015 to June 2017), evaporation summed 2843 mm, and rainfall 2524 mm, i.e., 89% of the losses due to evaporation, i.e., the rainfall over the reservoir almost covered the evaporation in the studied period.

From a regional perspective, another approach to the sustainability of the Funil HPP's WFP can be conducted. In this case, the HPP reservoirs in the Upper Grande River cascade regulate downstream flows, which increase water security, especially in the dry season. Thus, the increased regional water availability in the Grande River basin offsets the water losses through evaporation for power generation. The accumulated evaporation from the Funil HPP reservoir between July 2015 and June 2017 was equivalent to an average of 1.6 $m^3 s^{-1}$, while the plant's average flow used in the turbines in the same period was 184 $m^3 s^{-1}$.

The WFP obtained for the Funil HPP is relatively low compared to estimates for other HPPs in Brazil and worldwide. This can be explained by (1) the mild climate of the region where the plant is located, which results in median evaporation rates; (2) the small size of the reservoir, which results in lower evaporation volumes; (3) the existence of an upstream flow regularization reservoir (Camargos HPP); (4) and the high installed power related to the flooded area. Therefore, given the above assessments, the WFP of the Funil HPP can be considered sustainable, with a low impact on the basin's water resources.

7 Conclusion

WFP has been considered as an important indicator of the efficiency and sustainability of the water resources used in hydropower generation. Gross and net WFPs of the HPP, located on the Grande River, southeastern Brazil, were estimated considering climate variables monitored by weather stations installed in the reservoir's neighborhood, temperature of the water, recorded by sensors, and the electrical energy produced by the facility, taking into account the operational data provided by the energy company that operates the facility. Gross WFP is defined as the ratio of evaporation from the reservoir and respective energy production, whereas net WFP is the ratio of evapotranspiration of the flooded area of the reservoir before its filling less the evaporation of the reservoir and respective energy production. The annual gross WFP of the HPP was 29 $m^3.GJ^{-1}$, while the net WFP was 13 $m^3.GJ^{-1}$, that is, a value 55% lower when adopting the latter methodology. Yet, results showed an efficiency of the water use for electrical energy generation in the HPP due to the small WFP, even when considering the gross WFP as a reference, compared to other studies conducted in Brazil and worldwide. It has been found that the WFP of the HPP may be considered sustainable because of the positive water budget between availability and demand for water resources in the basin where the plant is installed.

Acknowledgments The authors thank CEMIG/ANEEL and FAPEMIG (CAG—APQ 03393-12) for financially supporting this research and scholarship for the first author. The authors also thank the "Conselho Nacional de Desenvolvimento Científico e Tecnológico (CNPq) for sponsoring the sabbatical research development (200207/2022-9) of the second author in the University of Illinois Urbana-Champaign, and Universidade Federal de Lavras for all support in this period.

References

Abtew W (2001) Evaporation estimation for Reservoir Okeechobee in South Florida. J Irrig Drain 127:140–147

Adegun O, Olalekan A, Badru G, Odunuga S (2018) Water, energy and agricultural landuse trends at Shiroro hydropower station and environs. Proc IAHS 376:35–43

Agência Nacional de Energia Elétrica (ANEEL) (2017) Banco de Informações de Geração – BIG. Brasília, 2017. Available on http://www2.aneel.gov.br/aplicacoes/capacidadebrasil/capacidadebrasil.cfm. Accessed 15 Dec 2017

Aguilar S, Louw K, Neville K (2011) IHA World Congress Bulletin. International Institute for Sustainable Development (IISD) and International Hydropower Association (IHA). Available on: www.iisd.ca/ymb/hydro/iha2011. Accessed 22 Jan 2018

Allen RG, Pereira LS, Raes D, Smith M (1998) Crop evapotranspiration: guidelines for computing crop water requirements. Roma: FAO. (Irrigation and Drainage Paper, 56). Available on http://www.fao.org/docrep/x0490e/x0490e00.htm. Accessed 10 Apr 2014

Bakken TH, Killingtveit Å, Engeland K, Alfredsen K, Harby A (2013) Water consumption from hydropower plants: review of published estimates and an assessment of the concept. Hydrol Earth Syst Sci 17:3983–4000

Bakken TH, Kjosavik F, Killingtveit A, Alfredsen K (2015) Are reservoirs water consumers or water collectors? Reflections on the water footprint concept applied on reservoirs. Water Resour Manage 29:4919–4926

Bakken TH, Modahl IS, Engeland K, Raadal HL, Arnoy S (2016) The life-cycle water footprint of two hydropower projects in Norway. J Clean Prod 113:241–250

Bakken TH, Killingtveit A, Alfredsen K (2017) The water footprint of hydropower production: state of the art and methodological challenges. Global Chall 1:1–8

Berger M, Finkbeiner M (2010) Water footprinting: how to address water use in life cycle assessment? Sustainability 2:919

BRASIL (2006) Ministério do Meio Ambiente. Caderno da Região Hidrográfica do Paraná. Brasília, 240 p.

Bueno EO (2018) Pegada hídrica das usinas hidrelétricas do sistema interligado nacional na região hidrográficas do Paraná. Thesis (Ph.D.), Federal University of Lavras (UFLA), Lavras
Bueno EO, Mello CR, Alves GJ (2016) Evaporation from Camargos hydropower plant reservoir: water footprint characterization. Rev Bras Recur Hidr 21:570–575
Bueno EO, Alves GJ, Mello CR (2020) Hydroelectricity water footprint in Parana Hydrograph Region, Brazil. Renew Energy 162:596–612
Coelho CD et al (2017) Comparison of the water footprint of two hydropower plants in the Tocantins River Basin of Brazil. J Clean Prod 153:164–175
De Bruin HAR (1982) Temperature and energy balance of a water reservoir determined from standard weather data of a land station. J Hydrol 59:261–274
Dias NL, Kan A (2008) Evaporação líquida no reservatório de foz do areia, PR: estimativas dos modelos de relação complementar versus balanço hídrico sazonal e balanço de energia. Rev Bras Recur Hidr 13:31–43
Dias NL, Rocha LS (1999) Cálculo da taxa de variação da entalpia para os lagos de Itaipu e Foz do Areia. RBRH - Revista Brasileira de Recursos Hídricos 4(3):39–51
Dias NL et al (2002) O Método de Covariâncias Turbulentas Atenuadas (MCTA) para medição dos fluxos de calor sensível e latente: aplicação ao lago de Itaipu e seu redor. Rev Bras Recur Hidr 7: 143–160
Elsawwaf M, Willems P, Feyen J (2010) Assessment of the sensitivity and prediction uncertainty of evaporation models applied to Nasser Reservoir, Egypt. J Hydrol 395:10–22
Finch JW (2001) A comparison between measured and modelled open water evaporation from a reservoir in south-east England. Hydrol Process 15:2771–2778
Gerbens-Leenes PW, Hoekstra AY, van der Meer TH (2009) The water footprint of energy from biomass: a quantitative assessment and consequences of an increasing share of bio-energy in energy supply. Ecol Econ 4:1052–1060
Giadrossich F, Niedda N, Cohen D, Pirastru M (2015) Evaporation in a Mediterranean environment by energy budget and Penman methods, Reservoir Baratz, Sardinia, Italy. Hydrol Earth Syst Sci Discuss 12:1901–1940
Grubert EA (2016) Water consumption from hydroelectricity in the United States. Adv Water Resour 96:88–94
Herath I, Deurer M, Horne D, Singh R, Clothier B (2011) The water footprint of hydroelectricity: a methodological comparison from a case study in New Zealand. J Clean Prod 19:1582–1589
Hoekstra AY (2017) Water footprint assessment: evolvement of a new research field. Water Resour Manage 31:3061–3081
Hoekstra AY, Chapagain AK, Aldaya MM, Mekonnen MM (2011) The water footprint assessment manual: setting the global standard. Earthscan, London, 224 p
Hogeboom RJ, Knook L, Hoekstra AY (2018) The blue water footprint of the world's artificial reservoirs for hydroelectricity, irrigation, residential and industrial water supply, flood protection, fishing and recreation. Adv Water Resour 113:285–294
INTERGOVERNMENTAL PANEL ON CLIMATE CHANGE – IPCC (2012) Special report on renewable energy sources and climate change mitigation. Cambridge University, Cambridge
INTERNATIONAL ENERGY AGENCY – IEA (2017) Electricity information: overview 2017. Available on www.iea.org/statistics/topics/electricity/. Accessed 15 Nov 2017
Liu J, Zhao D, Gerbens-Leenes PW, Guan D (2015) China's rising hydropower demand challenges water sector. Sci Rep 5:11446
McJannet DL, Webster I, Stenson M, Sherman BS (2008) Estimating open water evaporation for the Murray-Darling Basin. CSIRO, Camberra, 50 p
McMahon TA, Peel MC, Lowe L, Srikanthan R, McVicar TR (2013) Estimating actual, potential, reference crop and pan evaporation using standard meteorological data: a pragmatic synthesis. Hydrol Earth Syst Sci 17:1331–1363
Mekonnen MM, Hoekstra AY (2011) The water footprint of electricity from hydropower. UNESCO-IHE; Institute for Water Education, Delft, 36p. (Value of Water Research Report Series, 51)

Mekonnen MM, Hoekstra AY (2012) The blue water footprint of electricity from hydropower. Hydrol Earth Syst Sci 16:179–187

Mekonnen MM, Gerbens-Leenes PW, Hoekstra AY (2015) The consumptive water footprint of electricity and heat: a global assessment. Environ Sci Water Res Technol 1:285–297

Mekonnen MM, Gerbens-Leenes PW, Hoekstra AY (2016) Future electricity: the challenge of reducing both carbon and water footprint. Sci Total Environ 569/570:1282–1288

Mello CR, Vieira NPA, Guzman JA, Viola MR, Beskow S, Alvarenga LA (2021) Climate change impacts on water resources of the largest hydropower plant reservoir in southeast Brazil. Water 13:1560

Monteith JL (1965) Evaporation and environment. Water in Plant – Evaporation and Environment 19:205–235

Morton FI (1983a) Operational estimatives of areal evapotranspiration and their significance to the science and practice of hydrology. J Hydrol 66:1–76

Morton FI (1983b) Operational estimates of Reservoir evaporation. J Hydrol 66:77–100

Morton FI (1986) Practical estimates of reservoir evaporation. J Clim Appl Meteorol 25:371–387

Morton FI, Ricard F, Fogarasi S (1985) Operational estimates of areal evapotranspiration and Reservoir evaporation – Program WREVAP. National Hydrology Research Institute, Ottawa. (NHRI Paper 24)

Operador Nacional do Sistema Elétrico (2004) Evaporações líquidas nas usinas hidrelétricas. Rio de Janeiro, 61 p.

Pasqualetti M, Kelley S (2008) The water costs of electricity in Arizona. Arizona Water Institute, [S. l.]

Penman HL (1948) Natural evaporation from open water, bare soil and grass. Proc R. Soc Lond Ser A Contain Pap Math Phys 193:120–145

Rosenberry DO (2007) Comparison of 15 evaporation methods applied to a small mountain lake in the northeastern USA. J Hydrol 340(3/4):149–166

Scherer L, Pfister S (2016) Global water footprint assessment of hydropower. Renew Energy 99: 711–720

Silva VO, Mello CR (2021) Meteorological droughts in part of southeastern Brazil: Understanding the last 100 years. An Acad Bras Cienc 93(Suppl. 4):e20201130. https://doi.org/10.1590/0001-3765202120201130

Strachan IB, Tremblay A, Pelletier L, Tardif S, Turpin C, Nugent KA (2016) Does the creation of a boreal hydroelectric reservoir result in a net change in evaporation? J Hydrol 540:886–899

Tanny J, Cohen S, Assouline S, Lange F, Grava A, Berger D, Teltch B, Parlange MB (2008) Evaporation from a small water reservoir: direct measurements and estimates. J Hydrol 351: 218–229

Tanny J, Cohen S, Berger D, Teltch B, Mekhmandarov Y, Bahar M, Katul GG, Assouline S (2011) Evaporation from a reservoir with fluctuating water level: correcting for limited fetch. J Hydrol 404:146–156

Tefferi MEA (2012) The effect of Ethiopian hydropower reservoirs on Blue Nile River flow regime. 2012. Thesis (Master Science) – University of Science and Technology

Torcellini P, Long N, Judkoff R (2004) Consumptive water use for U.S. power production. National Renewable Energy Laboratory, Golden

Wilson W, Leipzig T, Griffiths-Sattenspiel B (2012) Burning our rivers: the water footprint of electricity. River Network, Portland

Yesuf MB (2012) Impacts of cascade hydropower plants on the flow of the river system and water level in Reservoir Turkana in Omo-Ghibe Catchment, Ethiopia. 2012. Thesis (Master Science) – Norwegian University of Science and Technology

Zhao D, Liu J (2015) A new approach to assessing the water footprint of hydroelectric power based on allocation of water footprints among reservoir ecosystem services. Phys Chem Earth 79/82: 40–46

Evaluation of the Water Footprint of the Use and Maintenance of Buildings. Application to Dwellings in Andalusia, Spain

Daniel R. Vaquedano-Ramírez, Patricia González-Vallejo, M. Rocio Ruiz-Perez, and Madelyn Marrero

1 Introduction

By carefully managing budgets and considering environmental impacts, construction companies can achieve both financial and environmental project success (Islam et al. 2015). This can involve a variety of strategies, such as the selection of materials that are more energy-efficient and/or exert a lower environmental impact (Alba-Rodríguez et al. 2022), the implementation of water conservation measures (Ruiz-Pérez et al. 2022) and the incorporation of renewable energy sources into the project design.

Effective management can also involve careful planning and forecasting, by closely monitoring costs and adjusting in order to stay within the allocated budget or within the bounds to attain environmentally sustainable goals (AbouHamad and Abu-Hamd 2019). In the particular case of water, conservation measures can help reduce the demand of water resources over the life cycle of the building.

In Spain, it is becoming more common for administrations, in order to grant building permits, to evaluate the environmental information about the project

D. R. Vaquedano-Ramírez · M. Marrero (✉)
Higher Technical School of Building Engineering, University of Seville, Seville, Spain
e-mail: madelyn@us.es

P. González-Vallejo
Higher Technical School of Engineering, University of Seville, Seville, Spain
e-mail: pgonzalez1@us.es

M. R. Ruiz-Perez
Polytechnical School of Engineering, University of Seville, Seville, Spain
e-mail: rruiz15@us.es

S. S. Muthu (ed.), *Sustainability and Water Footprint*, Environmental Footprints and Eco-design of Products and Processes, https://doi.org/10.1007/978-3-031-70810-7_5

involved. This began with the implementation of energy efficiency criteria in the construction of buildings (CTE 2006), and the control, management and recycling of construction and demolition waste (CDW) (RD 105/2008). In 2017, the law on public sector contracts incorporated sustainability criteria into the evaluation process for the awarding building projects in public tenders (Ministry of the Presidency of the Government of Spain 2017). However, its implementation remains difficult since this inclusion represents a paradigm shift in the sector. These criteria make it necessary to measure environmental impacts through different indicators with the help of computational tools in order to automate processes and provide information to users (Cai et al. 2022).

Another important aspect of buildings is that their useful life can last for several decades. Maintenance costs can account for a significant proportion of the total budget over the life cycle of the building. To minimise these costs, it is important to not only consider the design and construction of the building with a long-term view, including the selection of durable materials and systems, but also plan an appropriate preventive maintenance program.

In this work, the environmental and economic assessment tool, PREDICE (Marrero et al. 2024), is employed, which reads the project bill of quantities (BoQ) in an exchange format .bc3 (FIEBDC 2022) and evaluates its environmental impact. In order to generate the BoQ of future maintenance projects, the work breakdown system (WBS) of the Andalusian Construction Cost Database (ACCD 2017) is utilised. 95 maintenance tasks and their cost units are defined in the present work. Two actual projects in Andalusia are evaluated, a 10-storey block of social housing and a two-storey single-family private dwelling.

The main objective of this work is to propose a methodology for the classification and structure of maintenance work by defining work items in different periods of the building's life cycle and linking them to the calculation of its water footprint (WF). As a secondary objective, the PREDICE tool is validated, through its generation of maintenance and renovations of actual projects. Furthermore, the stages of the building's life that generate the greatest impacts are identified. The third and last objective is to obtain the direct WF, from the occupant utility consumption, and compare the direct and indirect impacts.

First, the WBSs employed in the construction sector are explored. They are employed to define the inventory of resources to be consumed during the building life cycle. The main information is obtained from the cost analysis or life cycle costing. Second, the evaluation tool PREDICE is employed for the simultaneous evaluation of cost and indirect WF. Also, a methodology is defined for the assessment of the direct WF in the building. Finally, the two case studies give results similar to those found in literature.

1.1 Construction Classification Systems

A major aspect regarding the incorporation of environmental impacts into project budgets can be found in the WBS or decomposition of the work. They break down a complex problem, the building construction, into smaller parts that can be combined to define a build environment. ISO-12006-2 organises and classifies the basic concepts of classification (ISO 2015). Among the internationally recognised systems, Omniclass (Omniclass 2012) and ICMS stand out (ICMS SSC 2019), since they contain life cycle classification criteria. While Omniclass contemplates classification tables according to the purpose, ICMS is focused on cost classification. In Spain, the construction cost databases have their own WBSs and are the most representative regionally oriented databases. These include: the Technological Institute of Construction of Catalonia (ITeC 2012); the BPCM Madrid (BPCM 2023); the BDC-IVE of Valencia (BDC-IVE 2023); the BDEU of the Basque Country (BDEU 2023); PRECIOCENTRO of Guadalajara (PRECIOCENTRO 2023); and the ACCD (Andalusia Construction Cost Database) in Andalusia (ACCD 2017). The Spanish Association of Construction Database Writers has striven to unify all systems by developing a standard exchange format, .bc3 (FIEBDC 2022).

In particular, the present work uses the ACCD, which is open and freely accessible and belongs to the area of analysis. Its most widespread use is for cost estimation in the construction of homes and its use is mandatory in public developments in Andalusia, Spain. The work units are divided into a categorised organisation where, at the top level, the entire project lies. At the top of the classification, groups are divided into subgroups of homogeneous characteristics. And, rising from lower to higher levels, groups accumulate the qualities of work units belonging to a lower group (Junta de Andalucía 2021).

The first level is made up of chapters that represent the processes involved in construction: Demolition, Earthworks, Foundations, Water Disposal, Structure, Partition, Roofing, Installation, Insulation, Finishes, Carpentry, Glass and Polyester, Finishes, Decoration, Urbanisation, Construction Safety and Waste Management. The next levels in this order are subchapters, sections and groups. The lowest level consists of the costs of the work units, which are organised into unit, auxiliary and basic costs.

The work units are formed by basic elements that define material, labour and machinery resources needed to execute the unit (see Fig. 1). Together, they provide a settled and very strong system (ACCD has been widely used in Andalusia for three decades) that secures the capability of the model (Junta de Andalucía 2021). The identification code is formed by letters and numbers, where "A" stands for alphabetic characters and "N" for numeric characters.

The classification system in the code has the following functions:

- It allows for the accurate detection of coded work units. This is due to the fact that the connection between the code and the unit is exclusive.
- It enables data management in digital environments.

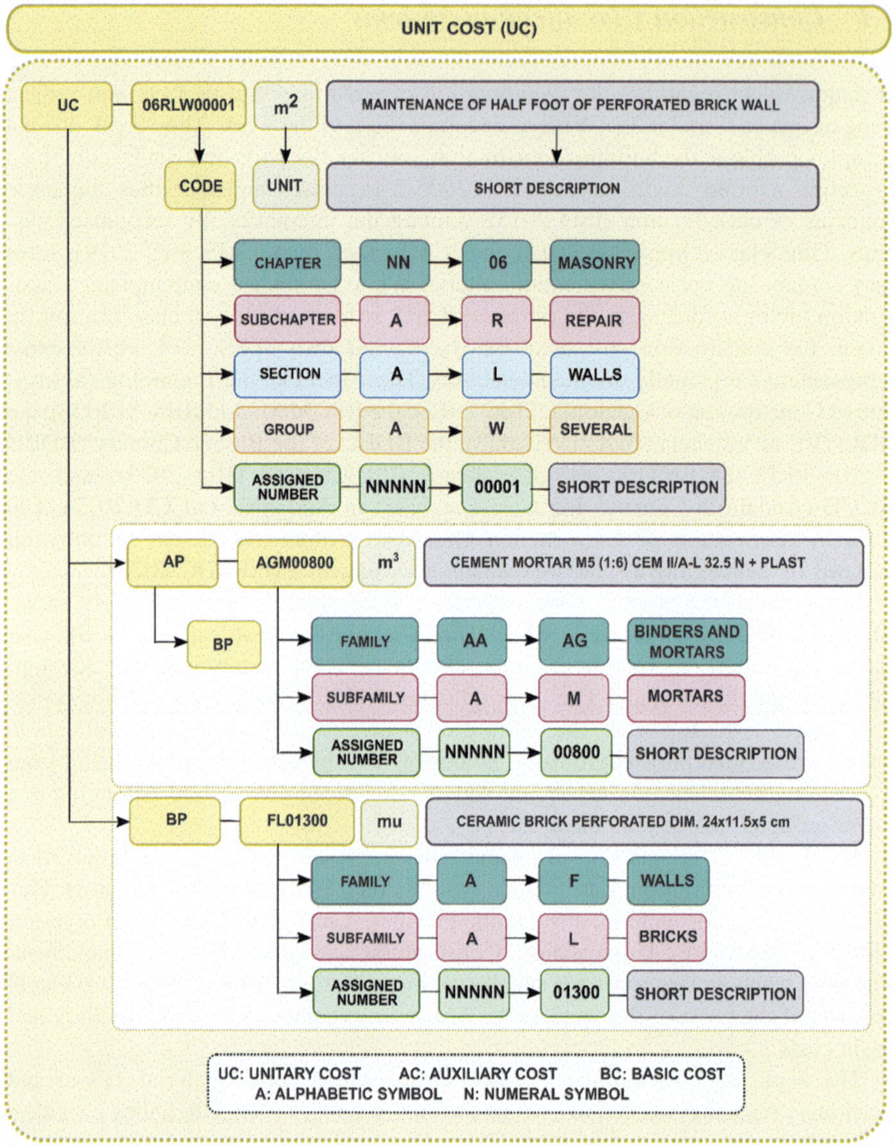

Fig. 1 ACCD coding hierarchy

- Resolves the place of items in the cost structure.
- It enables data exchange.

The description process necessitates a detailed explanation of each work unit and the resources necessary to complete a unit, which represents a finished element in the building. The resources (materials, machinery and labour) are the smallest part of the

Table 1 Example of the composition of a unit cost. Based on data from (ACCD 2017)

10SCS90040	**m^2**	**Porcelain ceramic floor with dimensions of 30 × 30 cm with adhesive**			
Flooring with 30 × 30 cm ceramic tiles, received with adhesive on a layer of M5 mortar (1:6), including levelling with a layer of sand of 2-cm thickness, a layer of mortar, smoothing paste, grouting and cleaning of the floor; built in accordance with the Spanish technical code.					
Measurement criteria: Measure the executed surface.					
Code	**Unit**	**Concept**	**Q.**	**BC(€)**	**Cost**
TO01100	h	Flooring officer	0.480	19.85	9.53
TP00100	h	Special worker	0.240	18.90	4.54
AA00200	m^3	Fine sand	0.020	12.92	0.26
AGL00100	m^3	Cement grout	0.001	116.28	0.12
AGM00500	m^3	Cement mortar M5	0.031	54.58	1.69
GP00100	Kg	Adhesive paste	1.000	0.22	0.22
GP00300	m^2	Levelling paste	1.000	1.55	1.55
RS02560	Unit	Ceramic tile 30 × 30 cm	11.780	2.64	31.10
			Direct cost		49.01

hierarchy. The work units descriptions also have a heading, unique, which includes the following elements: a unit of measurement, simple name, the resources, the construction task description, standards and/or regulations applicable and the way to measurement the work unit. Furthermore, shared measurement criteria are established for related work units. All these features together enable the inclusion of the environmental impact based on the same assumptions and boundaries as those defined in the estimate of the unit cost. Table 1 explains the unit cost of terrazzo flooring, with the measurement criterion being m^2 of floor installed.

The code of each work unit is organised on four levels (Barón et al. 2017):

- Chapters are defined by two numeric characters (10. Finishes).
- Subchapters are defined by three characters: two numeric, which determine the chapter, and one alphabetic, usually the first letter of the name (10S, Finishes, Floor).
- Sections are defined by four characters: two numeric and two alphabetic, the second of these being the first letter of the name of the section (10SC. Finishes, Floor, Ceramic).
- Groups are defined by five characters: two numeric and three alphabetic, the third of these being the first letter of the group name (10SCS. Finishes, Floor, Ceramic, Floor tile).

Finally, the unit cost is created,

- Headings are defined by 10 characters, two numeric, three alphabetic and five numeric at the end of the code. The latter will make it possible to allocate 100,000 items within each group, providing the coding with sufficient capacity to cover the planned objectives. (10SCS90040). See Table 1.

1.2 Life Cycle Costs

"Life cycle costing is concerned with optimising value for money in the ownership of physical assets by taking into consideration all the cost factors relating to the asset during its operational life" (Woodward 1997). It makes it necessary to assess the product cost and/or the economic effort made in its production, but extended across its useful life, incorporating its maintenance and disposal.

Life cycle cost (LCC) assessment began in the military industry (Sherif and Kolarik 1981). The phases included research and development, design, manufacturing, installation, operation, maintenance and recycling of parts. Throughout this analysis, the costs of the product over its useful life were considered. In Spain, in 2011, regulations in this regard were established in the acquisition of materials (Pastor Sánchez 2015).

The LCC concept was applied in construction by the end of the 1970s. However, it was not until 2006 that the International Organization for Standardization (ISO), through the technical committee ISO/TC 59, Building Construction, subcommittee SC14, prepared a standard, with its latest update in 2017 (ISO/TC 59/SC 142017). This standard describes the concept of whole life cycle costs (WLCCs), which is defined as "all the relevant, initial, future costs and benefits of an asset, throughout its useful life, while complying with functional requirements" (see Fig. 2). There are four stages in the life cycle of a building: construction, operation, maintenance and demolition.

The year 2021 witnessed the publication of EN *15643–1:2021, Sustainability of construction works—Framework for assessment of buildings and civil engineering*

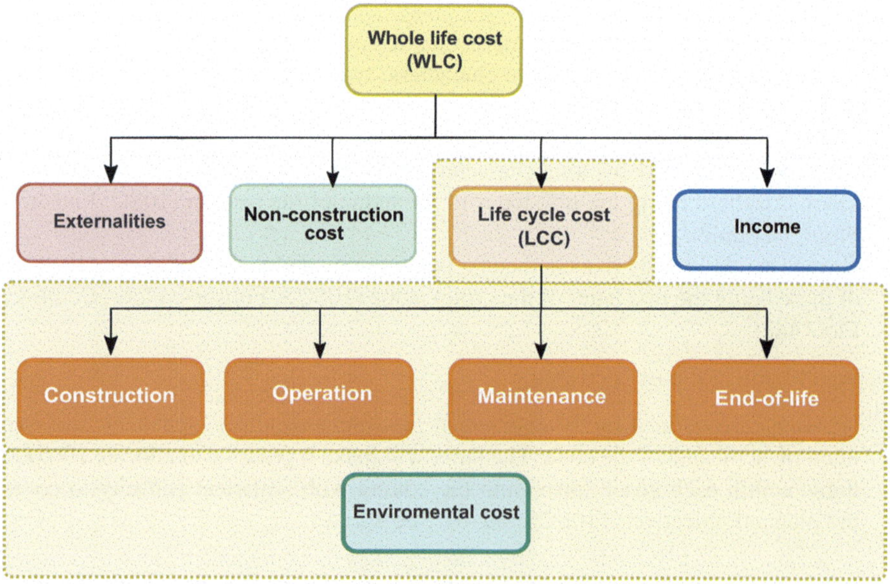

Fig. 2 Whole life cycle costs in buildings. Based on ISO/TC 59/SC 14 2017 and EN 15643–1:2021

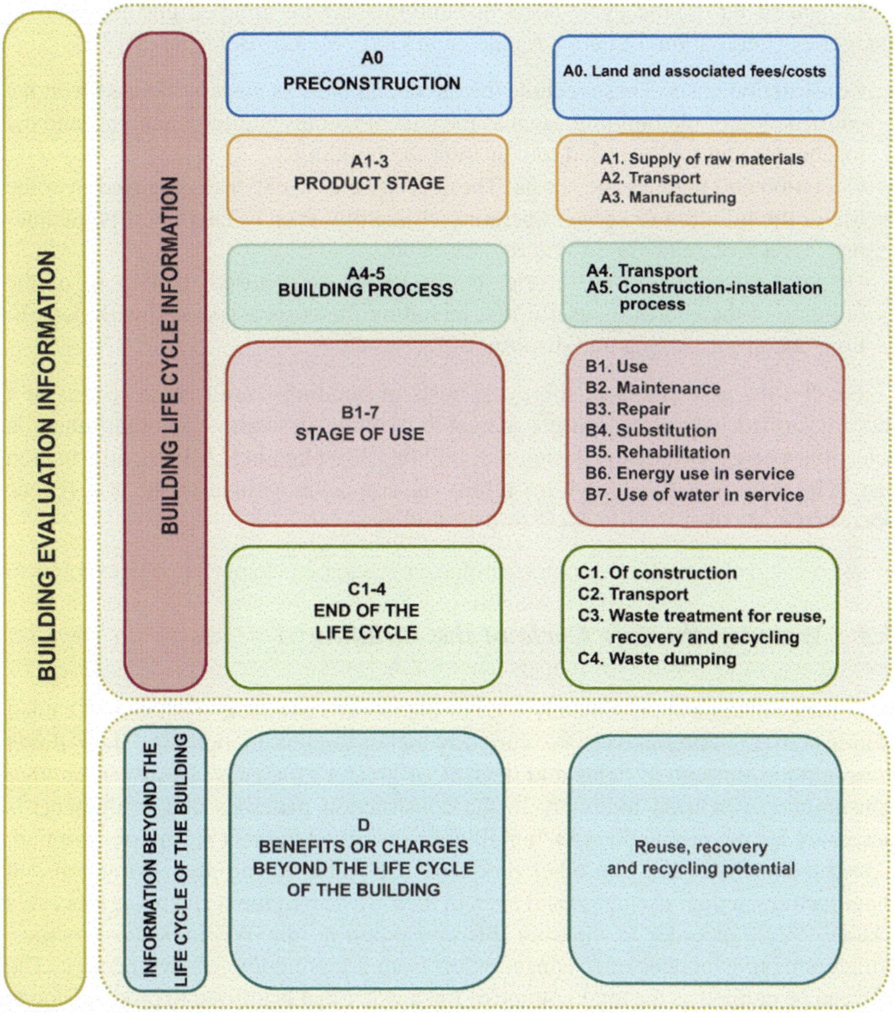

Fig. 3 Sustainability on construction sites. Evaluation of the economic performance of buildings. (Based on CEN/TC 350—Sustainability of construction works 2015)

works (see Fig. 3), which defines the regulatory framework and calculation method (TGO—Sustainability of construction works 2021). These established costs and revenues in the different stages of the building's life cycle: pre-construction, product stage, construction process, use stage and end of life cycle. It additionally includes a section that refers to benefits or costs beyond the life cycle of the building (its potential for reuse, recovery and recycling), which can be visualised in Fig. 3. Specifically, the standard is aimed at supporting both the decision-making process and the necessary documentation for the evaluation of building performance.

In general terms, life cycle costs in buildings can be grouped into three broad categories (International Energy Agency and ONU Medio Ambiente 2018):

- Construction costs: These include the direct and indirect costs associated with the construction of the building, such as the cost of labour, building materials and the machinery and tools needed to construct the building.
- Operation and maintenance costs: These costs refer to expenses incurred over the life of the building to keep it operating efficiently, such as energy costs, maintenance, repairs, renovations and replacements.
- Disposal costs: These costs refer to the expenses incurred to dispose of the building at the end of its useful life, including the costs associated with demolition, decommissioning and disposal of the waste.

Finally, the assessment of life cycle costs in buildings results in decisions of a more informed nature regarding sustainable building strategies, including material selection, energy efficiency, design and building life planning. All this information can help maximise the long-term return on construction investment and reduce operating costs over the life cycle of the building.

1.3 Water in the Life Cycle of the Building

Buildings consume approximately 30% of the world's available fresh water (United Nations 2012). Alternatives are currently being sought to optimise their direct consumption through systems and devices of greater efficiency. However, another consumption (incurred indirectly in the construction materials and equipment) is necessary for the production of a building: this is called indirect water consumption. Construction materials, like other products, use water during their extraction and manufacture, in their use phase and even in their waste treatment and reuse processes (Allan 1993). In order to measure this consumption, the WF indicator emerges, which aims to evaluate water consumption from a more complete perspective. The WF can be defined as the total volume of fresh water that is utilised to produce goods and services that are consumed by an individual or a community (Hoekstra and Hung 2002). The unit of measurement for this indicator is m^3, and it accounts for the volume of fresh water used directly and indirectly. There is also standardisation for its evaluation in that ISO 14046 (ISO 2014) establishes the requirements and guidelines necessary for the evaluation of products, processes and organisations, through the use of LCA (life cycle assessment). Another important aspect of the indicator in the construction sector is that it is easy to understand by all those involved, and it facilitates the study of a variety of products, materials and waste involved in the construction process with a clear message to society in the same way as other indicators that belong to the footprint family (Bovea and Powell 2016).

Very little research uses the WF indicator for the evaluation of direct and indirect water in buildings. Among the first studies on this topic is the evaluation of residential building construction in Australia and its life cycle (Crawford and Pullen

2011). These authors analysed water consumption in the life cycle of residential buildings over a 50-year period and concluded that the consumption of indirect water in building materials is higher than the direct consumption by occupants. In India, the WF of housing construction was determined (Bardhan 2011). In Iran, the WF of residential building construction was also assessed (Heravi and Abdolvand 2019), while in China, the construction of skyscrapers and dwellings was studied in terms of their WF (Meng et al. 2014; Han et al. 2016).

More recently in Spain, Rivero-Camacho and Marrero studied the WF indicator in the complete life cycle of residential buildings (Rivero-Camacho and Marrero 2022), and Ruiz-Pérez et al. considered urban service projects (Ruiz-Pérez et al. 2022). Lopes et al. have also studied 87 industrial projects to determine their WF (Lopes R. Silva et al. 2022). The latest studies have established a systematic classification of the on-site consumption of this resource, which enables its impact to be evaluated and shows the versatility of the indicator in the evaluation of building projects.

1.4 Environmental Impact Assessment Tools

Several tools that consider water consumption in buildings are available for the environmental assessment of construction projects, the most widespread of which are the Building Research Establishment Environmental Assessment Methodology (BREEAM 2017) and Leadership in Energy & Environmental Design (SpainGBC LEED 2017). The main obstacles as to why these tools are not widely used include their high cost, due to their mandatory licence fees and the great amount of work and expertise involved in carrying out the evaluation.

The ArDiTec research group, with whom we collaborate in the present work, is dedicated to the calculation of building indicators, and to their incorporation into construction cost databases (Freire-Guerrero et al. 2019). This methodology has been incorporated into a first computer tool called OERCO2, which is an online and open-source software for the calculation of the carbon footprint in the construction of new residential buildings (Solís-Guzmán et al. 2018). The user-friendliness of this software allows it to be employed by technicians with only basic knowledge and without the need for expertise in the field. Another tool is the Andalusian Ecological Certificate for the Certification of Buildings (CEACE), which certifies the sustainability of building construction and encompasses the ecological, economic and social aspects (González-Vallejo et al. 2020).

The CEACE has evolved into the PREDICE tool, which includes a more detailed calculation of the impacts of projects and directly reads the project's BoQ in an exchange format (Marrero et al. 2024). The calculations employ the WBS of ACCD, which means that each activity in the BoQ generates water, carbon and ecological footprints, in addition to embodied energy and CDW (see Fig. 4). The assessment covers the entire process, from the extraction of raw materials, manufacture, commissioning, maintenance and renovation to, lastly, demolition. The

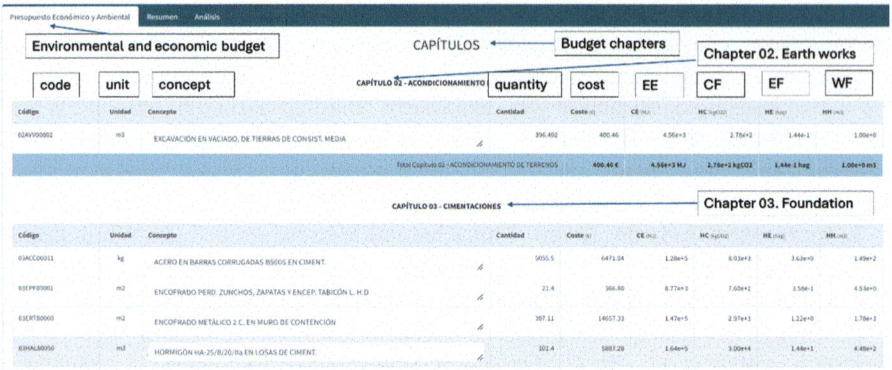

Fig. 4 Interface of the PREDICE tool, where the chapters of Sewers and Structures with their corresponding unit costs are shown, together with the economic and environmental impacts of each resource

environmental data has been obtained from the Ecoinvent database (Frischknecht et al. 2005). The internal calculations of the tool are explained in the following sections.

2 Methodology

The methodology for the calculation of the WF is shown in Fig. 5, where it is classified into several levels (sources of impact and footprints). It is based on an adjustment of the carbon and ecological footprint calculation model for buildings (Solís-Guzmán et al. 2013). The model has a cost structure, where each cost element has been ascribed impact coefficients in order to calculate its environmental impacts. It includes the indirect consumption of water, the embodied water in the fabrication of the construction materials, in the machinery fuel and in electricity production. The amount of water consumption by the building occupants and the water incorporated in their power consumption is also determined. The direct consumption by occupants in the form of tap water and electricity is obtained from statistical and reported data.

2.1 Indirect Consumption

It is necessary to establish a secondary database which allows the organisation of the substantial volume of data of the project, whether it be at different stages of the building life cycle, of construction, of use and maintenance or of demolition. It is required to develop a project management system, which organises the data necessary for the calculations during each stage of the lifecycle. The data is obtained from

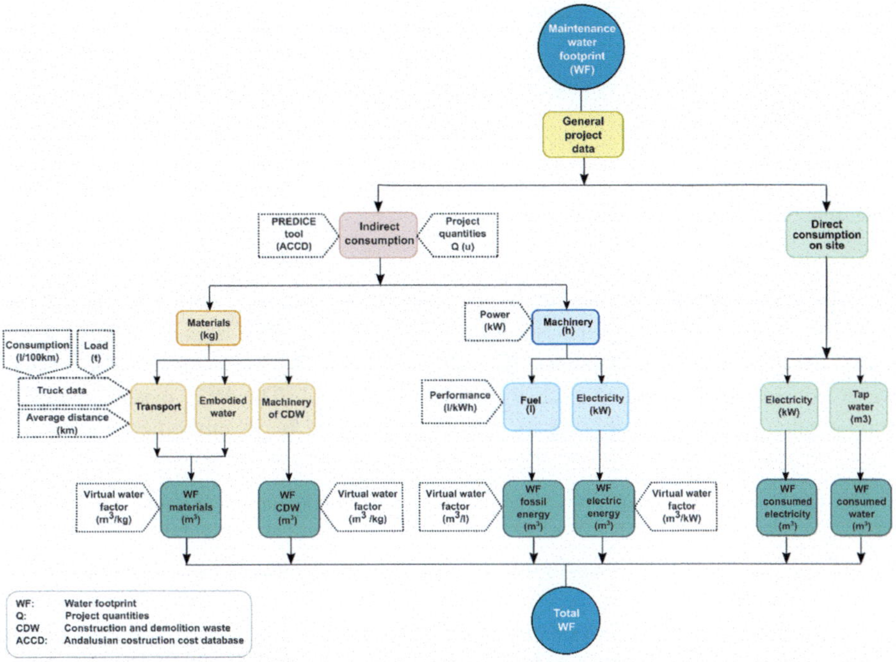

Fig. 5 WF methodology

the BoQ, which is structured in accordance with the WBS of the cost database. This, together with the WF methodology, has allowed the making of a data bank that combines environmental and cost data.

2.1.1 Materials

In order to obtain the WF of each element or the basic cost, it is necessary to determine the volume and weight of the product. Products can be sold by the cubic meter (m^3), square meter (m^2), meter (m), ton, thousands of units, etc. However, they need to be converted into cubic meters (m^3) so that the density specified in the supporting documents, such as the Catalogue of Construction Solutions of the Technical Building Code (CTE 2006), can be applied. This allows for the determination of the mass in kilograms (kg) for each basic element.

Once the WF has been determined for each of the construction materials per weight, their "environmental costs" are expressed per reference unit (its common reference units in the market, i.e. m^3 of concrete, tonne of steel, m of tube, etc.), which will generate the basic environmental impacts. The basic impacts created and the pyramidal structure allow for the calculation of unit impacts in the same manner as economic costs: by multiplying the quantities of each basic element by its unit impact. Hence, the new database of economic and environmental costs is created.

Table 2 Data for calculating the impact of transport

	Concrete	Other materials
Truck load capacity (kg)	24,000	2000
Distance to factory (km)	20	250
Average diesel consumption (litre/100 km)	26	26
Diesel emissions (tCO_2/litre)	2.62E-03	2.62E-03

Data from Freire-Guerrero et al. (2019)

Materials are classified by characteristics or elements of analogous physical properties. These are analysed by the Simapro computer tool (PRé Sustainability 2016), using Ecoinvent's ACL database (Ecoinvent Centre 2013), to obtain the virtual water consumed to produce each kilogram of material. The WF of each material can be calculated by multiplying its consumed material by its embodied water:

$$\mathrm{WF}_{\mathrm{ma}} = \sum (C_{\mathrm{ma}_i} \cdot \mathrm{VW}_{\mathrm{ma}_i}) \tag{1}$$

where C_{ma_i} denotes the consumption of material i (kg) and $\mathrm{VW}_{\mathrm{ma}_i}$ denotes virtual water of material i (m^3/kg).

The environmental impact of materials is assessed from cradle to gate stages. In order to evaluate the A4 aspect of UNE-EN 15978 (UNE-EN 15978:2012 2012), it is necessary to establish approximations of the distance travelled per transportation, see Table 2.

The WF of the transportation of materials is calculated as follows:

$$\mathrm{WF}_{\mathrm{tr}} = \sum \left(\frac{C_{\mathrm{ma}_i} W_{\mathrm{ma}i}}{T_{\mathrm{cap}}} \cdot D_{\mathrm{ma}} \right) \cdot T_{\mathrm{con}} \cdot \mathrm{VW}_{\mathrm{f}} \tag{2}$$

where $W_{\mathrm{ma}i}$ is the weight consumption of material i (kg); T_{cap} is the truck capacity (kg); D_{ma} is the average distance (km); T_{con} is the truck consumption (L/100 km); and VW_{f} is the virtual water of fuel (m^3/L).

2.1.2 Machinery

The VW of machinery is calculated using the power and performance of its engine and its working hours. The embodied water in a fuel litre consumed is different in diesel and petrol.

The equation for its calculation is,

$$\mathrm{WF}_{\mathrm{mc}} = \sum (H_{\mathrm{mc}_i} \cdot C_{f_i} \cdot \mathrm{VW}_{f_i}) \tag{3}$$

where H_{mc_i} denotes the working hours of machinery i (h); $C_{f\mathrm{i}}$ denotes consumption factor i (L/h or kW); and $E_{f\mathrm{i}}$ denotes virtual water factor of fuel i (m^3/L or m^3/kWh).

The power consumption of the electrical machinery or total KWh is determined by its engine power and its total working hours. The embodied water in a kWh generated by the Spanish energy mix is 0.0167 $m^3{}_{water}$/kWh (REE 2014).

2.1.3 Construction and Demolition Waste

CDW management activities are also included in the project BoQ, in compliance with Royal Decree 105/2008 (Royal Decree 105/2008 2008). The CDW impact is due to the pick-up and transport of waste to the landfill. For its calculation, Eq. 3 is used in the same way as for any other machine on the construction site.

2.2 PREDICE Tool

The methodology described above for the assessment of the indirect impact is incorporated into the PREDICE tool. In this chapter, the tool is tested for the first time with real cases. In the two case studies, the budget is generated in the exchange format .bc3. The new budgets are input into PREDICE to generate a table of the environmental outcomes. When the input is carried out, all the items must be verified as being valid: the software indicates recognised items in green, and those that do not exist in the database are indicated in red.

2.3 Direct Consumption by Occupants

2.3.1 Water

The Spanish National Institute of Statistics provides data on the average water consumption per person per year, from 1996 to 2015 (Instituto Nacional de Estadística 2019). A forecast is made that, by 2038, the WHO goal of 50 L/capita/day will be reached (Rivero-Camacho et al. 2023). This consumption is set as constant for the following years of the life cycle. The number of residents per home is obtained from the Spanish Technical Building Code (CTE 2006). The water consumption is 68.50 $m^3{}_{water}/m^2{}_{floor\ area}$ during a 100-year life span of the building (Rivero-Camacho and Marrero 2022). The life expectancy of the building is set at 100 years in order to meet the requirements of reinforced concrete structures in Spain (EHE 2008).

Other impacts associated with tap water include the energy consumed during its transport and losses due to leaks and breakdowns. In Spain, network losses 16% (INE 2019). Finally, the WF is 79.39 $m^3{}_{water}/m^2{}_{floor\ area}$.

2.3.2 Power Consumption

The energy consumption of occupants during the use stage changes over time due to modifications to the building envelope and its installations. The average electricity consumption of homes obtained from software simulation is 2713.37 kWh/ $m^2_{floor\ area}$ over 100 years. The WF of electricity is 0.0167 m^3/kWh (IDAE 2011), and hence the WF is 45.31 $m^3/m^2_{floor\ area}$.

2.4 Classification of Maintenance Work

The future maintenance tasks required by the project work units are defined. The compositions of all the work units and new codes are set, which are then translated into an interchange format (.bc3) and then added to the ACCD. Within the ACCD, maintenance items are defined in subchapter R. The tasks in turn are divided into different categories depending on the chapter, which determines the second letter of the code, and finally the third letter is established in a more specific way depending on the area in which it is performed. Table 3 shows the classification of repairs and replacements in chapter 5, Structures.

Table 3 ACCD classification of maintenance tasks in chapter 5, Structures, subchapter 5R. Repairs and replacements

Section		Group	
05RA	Steel	05RCA	Recesses
		05RAL	Cleaning and treatments
		05RAR	Reinforcements and consolidations
		05RAS	Replacements
		05RAW	Miscellaneous
05RF	Slabs	05RFA	Recesses
		05RFL	Cleaning and treatments
		05RFR	Reinforcements and consolidations
		05RFS	Replacements
		05RFW	Miscellaneous
05RH	Concrete	05RHA	Recesses
		05RHL	Cleaning and treatments
		05RHR	Reinforcements and consolidations
		05RHS	Replacements
		05RHW	Miscellaneous
05RM	Wood	05RMA	Recesses
		05RML	Cleaning and treatments
		05RMR	Reinforcements and consolidations
		05RMS	Replacements
		05RMW	Miscellaneous
05RW	Several	05RWA	Recesses
		05RWL	Cleaning and treatments
		05RWR	Reinforcements and consolidations
		05RWS	Replacements
		05RWW	Miscellaneous

Based on data from Junta de Andalucía (2021)

Table 4 Example of maintenance cost

05RFA90003	**m^2**	**Repair of a collaborative slab of 20-cm thickness**			
Maintenance of a collaborative slab of 20-cm thickness, formed by a steel sheet of 0.75-cm thickness, reinforced concrete HA-25/P/20/IIa with plastic consistency and maximum aggregate size 20 mm, reinforced with B500S steel, including a proportional part of solidifying supports, complementary formwork, shoring, stripping, vibrating and curing; built according to EHE, NCSR-02 and CTE. Measurement criteria: Measure the surface from outside-to-outside deducting gaps larger than 1 m^2.					
Code	**Unit**	**Concept**	**Quantity**	**€/unit**	**Cost (€)**
TO00600	h	Electrician	0.05	19.85	0.99
TO02100	h	Officer 1	0.08	19.85	1.59
TP00100	h	Special worker	0.50	18.90	9.45
CA00320	Kg	Steel B500S	3.00	0.81	2.43
CA00750	m^2	Collaborating steel sheet of 0.75 cm thickness	1.01	27.10	27.37
CH02920	m^3	Concrete HA-25/P/20/IIa	0.21	60.26	12.65
MV00100	h	Vibrator	0.15	1.51	0.23
WW00400	Unit	Small material	2.00	0.30	0.60
			Cost		**55.31**

Table 4 shows an example of a new maintenance unit cost. All the new codes proposed must be contained within the tree of the ACCD structure so that the new database can be imported into the PREDICE tool.

The significant years of the life cycle of the building are taken at 20, 40 and 70 years, as in previous work by the authors (Rivero-Camacho et al. 2023). The baseline study indicates which activities will need attention and at which stage of the life cycle.

3 Case Studies

The case studies include a two-storey single-family dwelling (P1) and a 10-storey multi-family building (P2), both located in the Andalusia area (see Fig. 6). Both are representative of the most frequent constructions in Spain (González-Vallejo et al. 2015).

The single-family dwelling consists of a ground floor, a second floor and a basement, as well as a garden and pool area. The foundation is made of reinforced concrete in footings, pile caps and retaining walls. The vertical structural elements are made of steel, and the slabs are made of reinforced concrete and a collaborating metal sheet. The roof is non-passable, consisting of a slope formation with M2.5

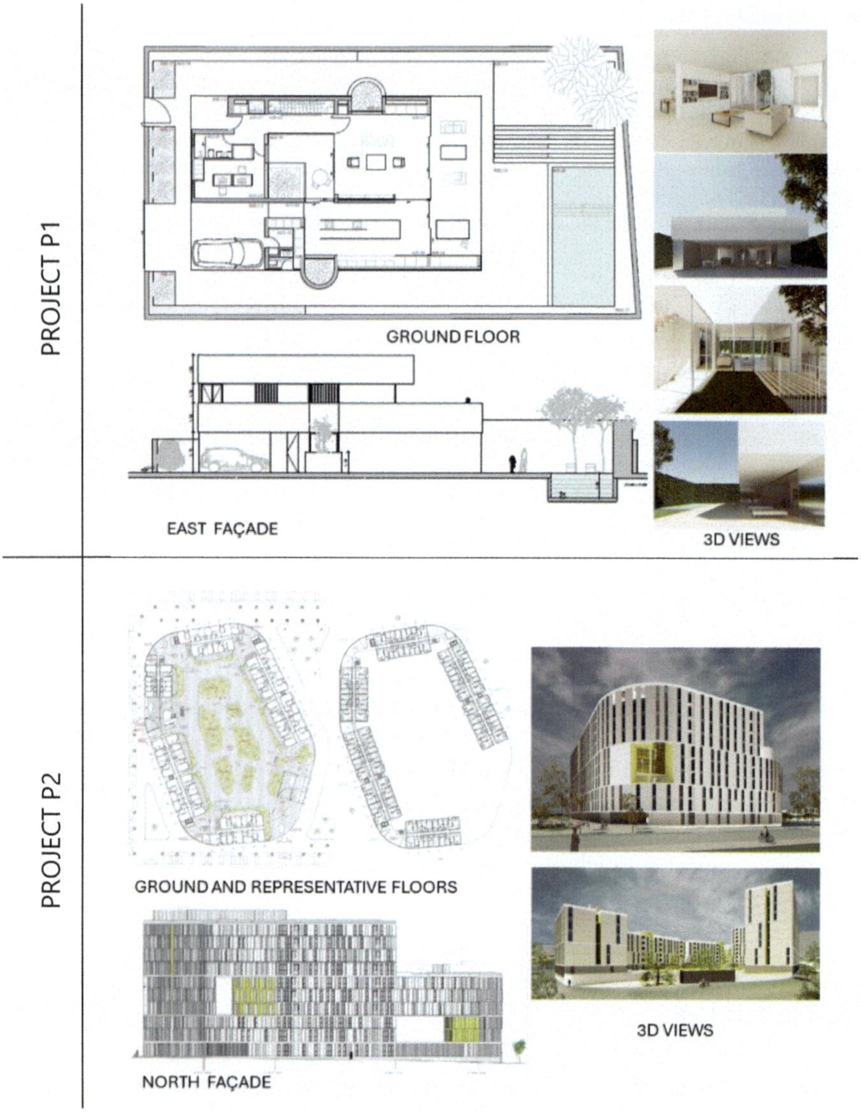

Fig. 6 Projects P1 and P2. (Based on project descriptions in JM Arquitectos (Vivienda [C], 2024), and Ma I Arquitecturas (218 Hacienda El Rosario – Ma I arquitecturas, 2024) respectively)

mortar, a bitumen membrane with double polyethylene mesh, a vapour diffuser layer, an extruded polystyrene insulation panel of 30-mm thickness and density of 25 kg/m^3, 100 g/m^2 polypropylene anti-punching shear fabric and a 5-cm-thick protective layer with rolled aggregate. The floor has 20 × 40 cm compact ceramic tiles on top of M5 mortar (1:6) and is levelled with a layer of sand. The dwelling floor area is 320 m^2.

The second project is a 10-storey multi-family building which has a reinforced concrete foundation. The structure is made of reinforced concrete; the roof is walkable and made of asphalt-based vapour barrier, with a 10-cm-thick layer of lightened concrete, a layer of mortar, a modified bitumen membrane of 4-mm thickness, with polyethylene reinforcement, a layer of protective mortar, and the flooring has ceramic tiles with dimensions of 14 × 28 cm on top of the mortar. The indoor flooring is mostly made up of 30 × 30-cm ceramic tiles, on a layer of M5 mortar, levelled with a layer of sand. The total construction area is 30,158 m^2, and a total of 218 dwellings are constructed.

3.1 Maintenance

It is necessary to define the maintenance work and prepare the maintenance BoQ for each period. All maintenance work units use the basic costs existing in the ACCD, which are, therefore, also held in PREDICE.

The activities established in each of the stages of maintenance are:

- Twenty-year renovation: This consists of the replacement of the air-conditioning, the heating and the domestic hot water (DHW) generation installations. Additionally, the façade is painted; although this is an activity that can be carried out more frequently, at approximately every 5 years according to certain studies (Martínez-Rocamora et al. 2016), for the sake of simplicity, only three maintenance periods are studied herein.
- Forty-year renovation: The second renovation includes the insulation of the roof and façades. The air-conditioning, the heating and the DHW installations are renovated again. Furthermore, the windows are replaced. The façade is plastered and painted. The roof is repaired. The old lifts or elevators are replaced with new ones.
- Seventy-year renovation: At this point in the life cycle, it is necessary to carry out an inspection and repair of structural elements such as fissures and cracks. A replacement of all electrical installations and sewer systems is also needed. A renovation of the air-conditioning and hot water installations is again carried out. The interior wall partitions, doors, kitchens and bathrooms are replaced. Structural repairs and painting are also carried out. Moreover, the renovation cost cannot exceed 60% of the construction cost of the original building in order to not be considered a new building in Spanish legislation (Miguel 2011).

In the present work, it has been necessary to create 48 maintenance tasks and unit costs for P1, and 95 tasks for P2.

4 Results

The proposed methodology provides the economic quantification, from which the environmental quantification of the life cycle of the building is attained for each maintenance stage. The results are fully obtained from the PREDICE tool, thanks to the input of the BoQ of the projects to automatically generate the environmental information of each phase. The final comparison between the two projects can provide information of the different stages depending on the building typology. Each of the items that form part of the budget generates an economic cost and WF. The results of each item are added together to obtain the results per chapter. Lastly, the sum of all the chapters of the budget provides the total impact of the project.

4.1 Construction

Starting with the construction stage of the projects, which corresponds to the beginning of the life cycle, the typologies of the materials used are taken into consideration, as are the construction techniques and the resources necessary for the manufacture of these materials.

The Economic results of the construction stage of both case studies are shown in Table 5, where P1 is the single-family dwelling project and P2 is the 10-storey

Table 5 Economic results of the construction phase

Chapters	P1 (€)	P2 (€)	P1 (€/m2)	P2 (€/m^2)
Demolition and preliminary work	–	8082.01	–	0.27
Land conditioning	400.46	69,285.00	1.25	2.30
Foundations	37,891.08	1,488,504.87	118.41	49.36
Sewers	11,299.11	138,893.80	35.31	4.61
Structures	73,201.27	2,058,488.17	228.75	68.25
Masonry	36,532.47	2,773,475.42	114.16	91.96
Roofs	10,206.07	286,196.14	31.89	9.49
Installations	106,674.59	3,921,284.18	333.36	130.02
Insulation	4000.64	860,024.70	12.50	28.52
Finishes	83,153.99	3,112,739.10	259.86	103.21
Carpentry and protection	28,544.17	1,237,562.50	89.20	41.30
Glasswork and synthetic products	17,912.67	156,960.31	55.98	5.20
Paints	4891.13	376,833.12	15.28	12.49
Equipment	51.56	23,764.37	0.16	0.79
Urbanisation	27,581.36	350,025.10	86.19	11.61
Waste management	3562.61	519,053.48	11.13	17.21
Health and safety	1731.38	–	5.41	–
Total	**447,634.56**	**17,381,172.26**	**1398.84**	**576.32**

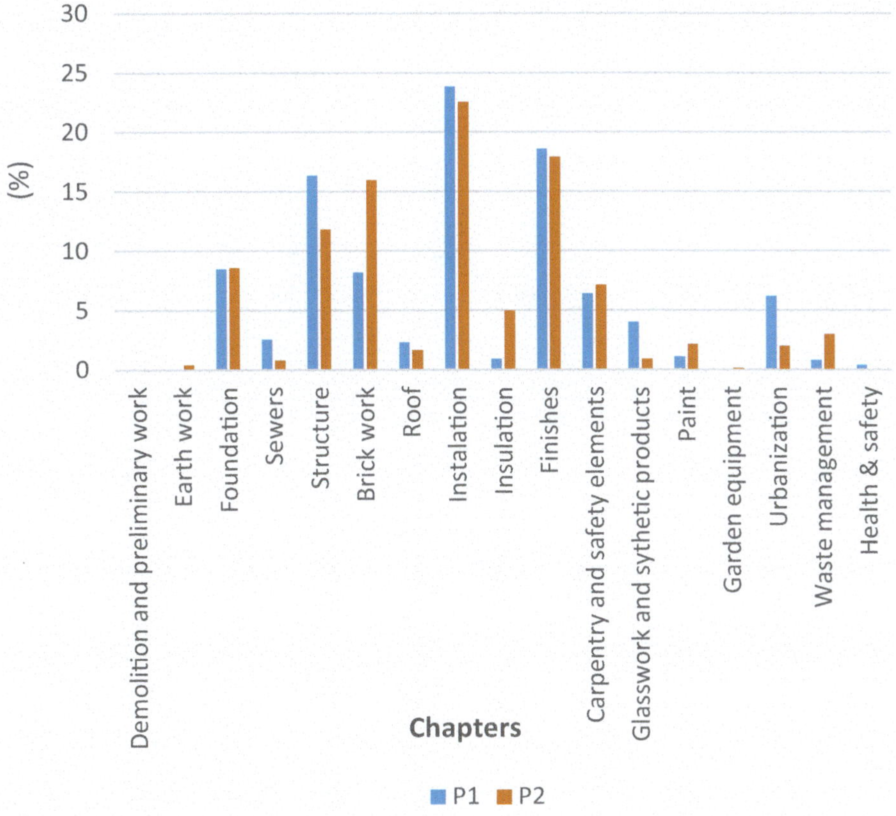

Fig. 7 Economic results by chapters under construction

building project. P1 is almost three times more expensive per square metre than is P2. Only the insulation chapter is more expensive in the multi-family housing due to the thermal demands of social housing. The big difference is due to the fact that the single-family home is private while the multi-family home is social housing, whereby, first, the installations and, second, the finishes are highly significant in both projects. The structure, foundation and masonry are also significant factors. These five chapters represent approximately 70% of the total cost of both projects.

Figure 7 shows the distribution in terms of percentages of costs, which are very similar in both projects. It can be observed that the costs of the installations per square metre are the most significant. In both projects, the distribution of costs is very similar, with installations, finishes and structures in P1 and masonry replacing structures in P2 as the most significant chapters. The difference in the structural impact is due to the greater use of steel in the single-family dwelling, whereas the multi-family dwelling uses more masonry per square metre of floor. The wall partitions are made of brick in both projects, and hence they are more numerous in social housing due to its smaller rooms, and its increased impact per square metre of floor.

Table 6 WF results in the construction stage

Chapters	WF P1 (m^3)	WF P2 (m^3)	WF P1 (m^3/m^2)	WF P2 ($m3/m^2$)
Demolition and preliminary work	0.00	37.32	0.000	0.001
Land conditioning	1.00	3510.88	0.003	0.116
Foundations	3045.12	131,868.95	9.516	4.372
Sewers	328.71	4638.19	1.027	0.154
Structures	1918.31	123,996.37	5.995	4.111
Masonry	557.11	88,539.49	1.741	2.936
Roofs	161.87	6517.89	0.506	0.216
Installations	1402.13	84,805.74	4.382	2.812
Insulation	18.28	3003.29	0.057	0.100
Finishes	1196.96	200,214.11	3.740	6.639
Carpentry and safety and security elements	756.32	8965.64	2.364	0.297
Glasswork and synthetic products	62.75	1272.93	0.196	0.042
Paints	129.65	16,629.75	0.405	0.551
Equipment	0.27	338.73	0.001	0.011
Urbanisation	966.52	10,837.48	3.020	0.359
Waste management	49.69	4658.34	0.155	0.154
Health and safety	125.30	0.00	0.392	0.000
Total	**10,720.00**	**689,835.10**	**33.500**	**22.873**

The P1 project has a metal structure, which is more expensive than reinforced concrete in P2. The construction of the swimming pool is included in the urbanisation of P1, which increases the importance of the chapter.

In the case of the WF results, Table 6 shows that in the P1 case study, the chapter that controls the footprint is that of the foundation, followed closely by the structure. These represent the elements with the highest water consumption in their manufacture, due to the intensive use of aggregates and stones, in addition to concrete and steel, which are also water-intensive materials. In the P2 case study, the finishes in the façade that is made of mortar have the largest WF, followed by the foundation. This also applies to flooring that includes stone material extracted from quarries, a process that is water-intensive. The total WF is 33.50 and 22.87 m^3/m^2 of floor area for single-family and multi-family dwellings, respectively. The distribution of the WF per chapter is similar to that found in the construction of 83 warehouse buildings in Seville (Lopes R. Silva et al. 2022), although these buildings have a lower WF due to the simplicity of their construction, at 6–12 m^3/m^2. Other authors results are 20.83 m^3/m^2 (Meng et al. 2014), 26.6 m^3/m^2 (Han et al. 2016) and 20.80 m^3/m^2 (Heravi and Abdolvand 2019). However, those projects fail to entail any detailed measurements for the comparison of the impacts per chapter or stage of the work.

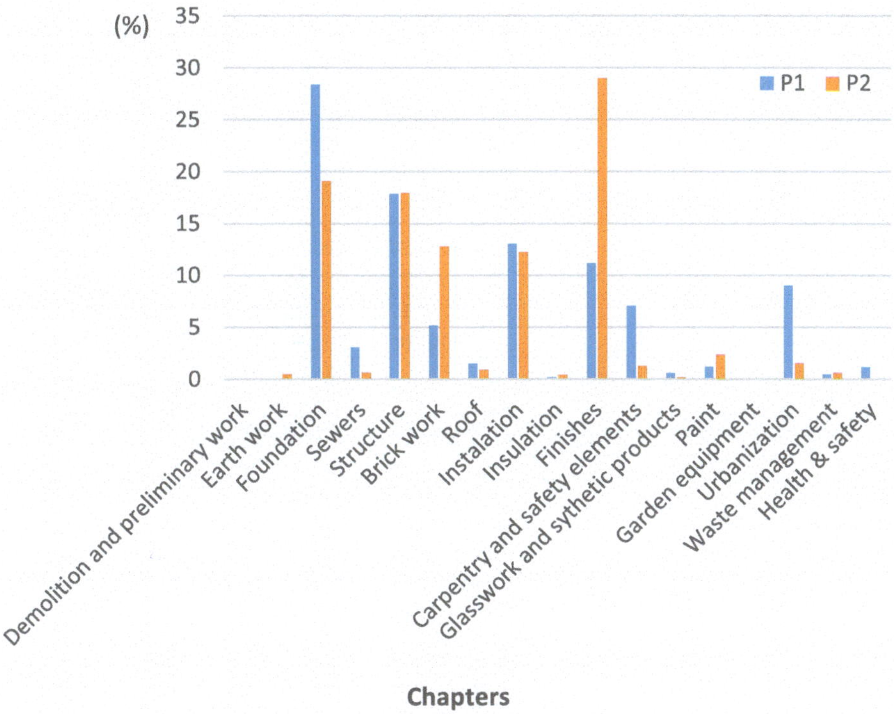

Fig. 8 WF results by chapters in construction budget

The slight differences may be due to the fact that, in the present work, the definition of the project is complete, and that almost 300 work units have been evaluated.

Figure 8 shows the WF results obtained in the construction stage of both projects in relation to the percentage of incidence of each chapter. The results lie within the usual ranges (Rivero-Camacho et al. 2023) since they also depend, to a large extent, on the size and complexity of the projects studied. This figure highlights the significance of the finishes chosen for the P2 project, which include stone used in the flooring, compared to the ceramic material used in P1. The foundation of P1 is more water-intensive due to its complexity since it is formed of retaining walls and piles.

At the end of the construction stage, the use stage begins, whereby maintenance must be carried out to ensure optimal functioning of the building. The economic results of the chapters that need maintenance and renovations per period can be visualised in Table 7.

For the two case studies, the first renovation involves only the Installations and Painting chapters, and therefore the investment is smaller compared to the costs of construction and other renovations, as shown in greater detail later. The results of the 20-year maintenance footprint can be found in Table 8, where the significant impact

Table 7 Economic results of the 20-year maintenance

Maintenance tasks	P1 (€)	P2 (€)	P1 (€/m^2)	P2 (€/m^2)
20 years				
Installations	1150.19	218,166.56	3.59	7.23
Paints		363,444.90		12.05
Total	**1150.19**	**581,611.46**	**3.59**	**19.28**
40 years				
Masonry	7904.78	420,184.81	24.7	13.93
Roofs	8397.72	287,651.88	26.24	9.54
Installations	1150.19	218,166.56	3.59	7.23
Insulation	3541.06	793,306.04	11.07	26.3
Carpentry and safety and protection elements		245,986.40		8.16
Glasswork and synthetic products	4077.06	181,199.22	12.74	6.01
Paints		363,444.90		12.05
Total	**25,070.81**	**2,509,939.81**	**78.34**	**83.22**
70 years				
Sewers	2140.44	18,477.80	6.69	0.61
Structure	25,494.13	1,902,870.29	79.67	63.09
Masonry	7904.78	420,184.81	24.70	13.93
Roof	2418.33	287,651.88	7.56	9.54
Installations	12,937.40	892,073.42	40.43	29.58
Finishes	39,484.09	1,483,184.57	123.39	49.18
Carpentry and safety and protection elements		245,986.40		8.16
Glasswork and synthetic products	4077.06	181,199.22	12.74	6.01
Paints		363,444.90		12.05
Total	**94,456.23**	**5,795,073.29**	**295.18**	**192.15**

of the paints can be appreciated. P1 does not need to be painted at the 20-year stage because its façade is covered with a single layer of mortar.

Renovation after 40 years involves more number of chapters. The WF values per square metre of the floor in the two case studies are relatively close to each other. The renovation that takes place after 70 years is the most comprehensive, compared to the previous two, resulting in a greater impact. Again, the impact is controlled by the WF of the structure, installations and finishes. The repairs to the structure are significant in terms of the WF due to the use of steel and concrete. These WFs account for 10% to 20% of the footprint of constructing a new building (see Fig. 9).

The Structure chapter has a higher economic and environmental impact, and therefore special control measures and improved design can significantly reduce the impact of the projects, since the repair work on the concrete structure is more significant.

The direct WF is 79.39 m^3/m^2 of floor area and the average electricity consumption of homes is 45.31 m^3/m^2 WF. Therefore, the WF from the direct consumption of

Table 8 Maintenance water footprint (WF)

Twenty years	WF P1 (m^3)	WF P2 (m^3)	WF P1 (m^3/m^2)	WF P2 (m^3/m^2)
Installations	9.80	2149.22	0.031	0.071
Paints		14,066.18		0.466
Total	**9.80**	**16,215.40**	**0.030**	**0.540**
40 years				
Masonry	100.98	4108.31	0.316	0.136
Roofs	183.17	6517.89	0.572	0.216
Installations	9.80	2149.22	0.031	0.071
Insulation	53.11	1457.95	0.166	0.048
Carpentry and safety and protection elements		1572.94		0.052
Glasswork and synthetic products	25.65	1272.80	0.080	0.042
Paints		14,066.18		0.466
Total	**372.71**	**31,145.28**	**1.165**	**1.031**
70 years				
Sewers	36.34	801.36	0.114	0.027
Structures	481.31	85,611.89	1.504	2.839
Masonry	100.98	4108.31	0.316	0.136
Roof	53.53	6517.89	0.167	0.216
Installations	395.58	19,605.64	1.236	0.650
Finishes	183.70	22,853.16	0.574	0.758
Carpentry and safety and protection elements	–	1572.94	–	0.052
Glasswork and synthetic products	25.65	1272.80	0.08	0.042
Paint	–	14,066.18	–	0.466
Total	**1277.09**	**156,410.18**	**3.990**	**5.186**

the building use, water and electricity represents approximately 80% of the impact; see Table 9. It is interesting to note that the water embodied in the electricity production of buildings is just as significant as the indirect water from materials and machinery employed in the construction and maintenance of the building. Therefore, special attention should be paid to this impact source. Making the building more energy-efficient would not only reduce its carbon footprint but would also significantly reduce its WF.

Stage	P1 WF (m^3/m^2)	P2 WF (m^3/m^2)
Construction	33.50	22.87
Renovation 20	0.03	0.54
Renovation 40	1.17	1.03
Renovation 70	3.99	5.19

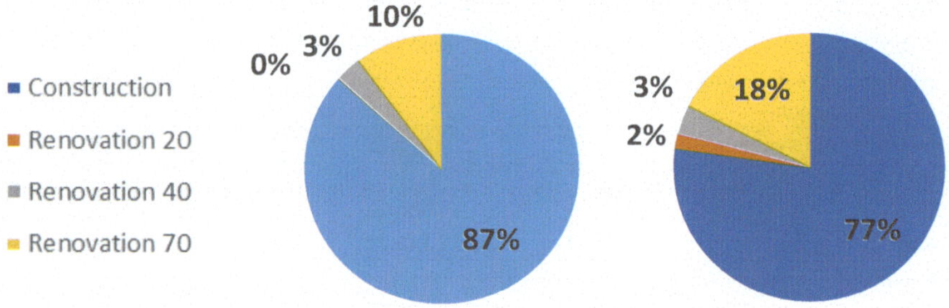

Fig. 9 Water footprint of the construction and maintenance of single-family housing P1 and multi-family housing P2

Table 9 Comparison of direct and indirect water footprint (WF) of the building's construction and maintenance

Source	P1 WF (m^3/m^2)	P2 WF (m^3/m^2)	% P1	% P2
Total indirect	38.69	29.63	23.68	19.20
Water utilities	79.39	79.39	48.59	51.44
Electricity utilities	45.31	45.31	27.73	29.36
Total WF	**249.78**	**240.72**	**100.00**	**100.00**

5 Conclusion

Environmental impact analysis of two types of building projects in Andalusia, a private single-family dwelling and a 10-storey multi-family block of social housing, was carried out throughout their life cycle. The construction phase was considered, as was the necessary maintenance after 20, 40 and 70 years. The construction stage and the maintenance when the building is 70 years old constitute the stages with the greatest economic impact and WF of the entire cycle.

For the economic and environmental evaluation, the user-friendly computer tool PREDICE is used. The tool works together with other tools, such as construction cost databases, in an easy and dynamic way. The tool allows the environmental aspect to be considered by professionals of the building sector and generates a greater awareness of the materials and construction methods employed.

The present methodology enables new maintenance and renovation work units to be defined based on the BoQ of the construction project. These new maintenance costs are easily inserted into the WBS of the construction cost database chosen. The tool is validated through the inclusion of these new maintenance costs, and PREDICE is able to detect these costs and generate results. In this way, it is possible to carry out an environmental analysis by means of including LCA data in the tool.

The results obtained for the two case studies are comparable with those of other similar projects. The WF is 33.50 and 22.87 m^3/m^2 in single-family and multi-family housing constructions, respectively. Among the results broken down by chapter, it is observed that the chapter on structural repairs has the greatest impact when the building is 70 years old: this is responsible for 10% and 18% of the total indirect WF, respectively. Other maintenance activities exert a small impact of less than 5%. However, the direct consumption by the buildings' occupants is the most significant factor, whereby the direct water and energy consumption represents approximately 80% of the total impact in both case studies.

Future research should focus on the use of environmental product declarations in order to ascertain and replace work units causing the greatest impacts with other commercial materials that cause less impact, instead of using the generic values produced by the PREDICE tool. The methodology can also be linked to project designs using Building Information Modelling applications. Furthermore, the energy simulation and actual water consumption values could provide a better comparison between the sources of impact.

References

218 Hacienda El Rosario – Ma I arquitecturas. https://mm-a.es/218-Hacienda-El-Rosario. Accessed 20 May 2024

AbouHamad M, Abu-Hamd M (2019) Framework for construction system selection based on life cycle cost and sustainability assessment. J Clean Prod 241:118397. https://doi.org/10.1016/j.jclepro.2019.118397

ACCD (2017) Andalusia construction cost database (ACCD). Andalusia Government. https://www.juntadeandalucia.es/organismos/fomentoyvivienda/areas/vivienda-rehabilitacion/planes-instrumentos/paginas/bcca-sept-2017.html

Alba-Rodríguez MD, Solís-Guzmán J, Marrero M (2022) Evaluation model of the economic-environmental impact on housing recovery. Application in the city of Seville, Spain. Sustain Cities Soc 83:103940. https://doi.org/10.1016/j.scs.2022.103940

Allan JA (1993) Fortunately, there are substitutes for water otherwise our hydro-political futures would be impossible. Priorities for water resources allocation and management

Bardhan S (2011) Assessment of water resource consumption in building construction in India. Ecosyst Sustain Dev 144:1743–3541. https://doi.org/10.2495/ECO110081

Barón L, Conde J, Osuna M, et al (2017) clasificación sistemática de precios básicos, auxiliares y unitarios

BDC-IVE (2023) Base de datos del Instituto Valenciano de Edificación (Base de datos del Instituto Valenciano de Edificación). In: Ministry of Infrastructure, Territory and Environment. https://productos.five.es/producto/base-de-datos-de-construccion-2023. Accessed 20 Nov 2023

BDEU (2023) Base de Precios de Edificación y Urbanización (Building and Urbanization Price Database). In: Gobierno Vasco.. http://www.euskadieuprecios.com/. Accessed 12 Mar 2023

Bovea MD, Powell JC (2016) Developments in life cycle assessment applied to evaluate the environmental performance of construction and demolition wastes. Waste Manag 50:151–172

BPCM (2023) Base de Precios de la Construcción (Construction Price Database), Madrid. In: Ministry of the Environment and Planning of the Territory. https://www.comunidad.madrid/servicios/vivienda/base-datos-construccion. Accessed 20 Nov 2023

BREEAM (2017) BREEAM. In: http://www.breeam.es/

Cai H, Wang X, Kim J-H et al (2022) Whole-building life-cycle analysis with a new GREET® tool: embodied greenhouse gas emissions and payback period of a LEED-certified library. Build Environ 209:108664. https://doi.org/10.1016/j.buildenv.2021.108664

CEN/TC 350 – Sustainability of construction works (2015) EN 16627:2015 Sustainability of construction works – Assessment of economic performance of buildings – Calculation methods

Crawford R, Pullen S (2011) Life cycle water analysis of a residential building and its occupants. Build Res Inf 39:589–602. https://doi.org/10.1080/09613218.2011.584212

CTE (2006) Spain MH (Ministry of Housing). Código Técnico de la Edificación (Building Technical Code). Madrid. Spain

Ecoinvent Centre (2013) Ecoinvent database v3 ecoinvent report

EHE (2008) REAL DECRETO 1247/2008, de 18 de julio, por el que se aprueba la instrucción de hormigón estructural (EHE-08). ROYAL DECREE 1247/2008, of July 18, which approves the structural concrete instruction (EHE-08)

FIEBDC (2022) Formato de Intercambio Estándar de Bases de Datos de Construcción. In: Asociación FIEBDC. https://www.fiebdc.es/. Accessed 12 Mar 2023

Freire-Guerrero A, Alba-Rodríguez MD, Marrero M (2019) A budget for the ecological footprint of buildings is possible: a case study using the dwelling construction cost database of Andalusia. Sustain Cities Soc 51:101737. https://doi.org/10.1016/j.scs.2019.101737

Frischknecht R, Jungbluth N, Althaus H-J et al (2005) The ecoinvent database: overview and methodological framework (7 pp). Int J Life Cycle Assess 10:3–9

González-Vallejo P, Solís-Guzmán J, Llácer R, Marrero M (2015) La construcción de edificios residenciales en España en el período 2007–2010 y su impacto según el indicador Huella Ecológica. Informes de la Construccion 67:e111. https://doi.org/10.3989/ic.14.017

González-Vallejo P, Alba-Rodríguez M, Rivero-Camacho C, et al (2020) Certificación de sostenibilidad en viviendas con la herramienta CEACE: Certificado Ecológico Andaluz para la Construcción de Edificios. Según indicadores ambientales Huella de Carbono (HC), Ecológica (HE) e Hídrica (HH). Congreso de Innovación y Sostenibilidad en la Vivienda Social

Han MY, Chen GQ, Meng J et al (2016) Virtual water accounting for a building construction engineering project with nine sub-projects: a case in E-town, Beijing. J Clean Prod 112:4691–4700. https://doi.org/10.1016/J.JCLEPRO.2015.07.048

Heravi G, Abdolvand MM (2019) Assessment of water consumption during production of material and construction phases of residential building projects. Sustain Cities Soc 51:101785. https://doi.org/10.1016/J.SCS.2019.101785

Hoekstra AY, Hung PQ (2002) Virtual water trade: A quantification of virtual water flows between nations in relation to international crop trade, value of water research report Series no:11

ICMS SSC (2019) International Construction Measurement Standards: Global Consistency in Presenting Construction Costs

IDAE (2011) Factores de conversión a Energía Primaria y factores de emisión de CO2 / CO2 emission factors

Instituto Nacional de Estadística (2019) National Institute of Statistics

International Energy Agency, ONU Medio Ambiente (2018) 2018 Informe Global Hacia un sector de edificios y de la construcción eficiente, resiliente y con cero emisiones

Islam H, Jollands M, Setunge S (2015) Life cycle assessment and life cycle cost implication of residential buildings—a review. Renew Sust Energ Rev 42:129–140. https://doi.org/10.1016/j.rser.2014.10.006

ISO (2014) ISO 14046:2014(es), Environmental management — Water footprint — Principles, requirements and guidelines. https://www.iso.org/obp/ui#iso:std:iso:14046:ed-1:v1:es. Accessed 5 May 2021
ISO (2015) ISO 12006-2 Building construction—Organization of information about construction Works
ISO/TC 59/SC 14 (2017) ISO 15686-5:2017: Buildings and constructed assets Service life planning Part 5: Life-cycle costing
ITeC (2012) Institute of Construction Technology of Catalunia
Junta de Andalucía (2021) Base de Costes de la Construcción de Andalucía (BCCA) – Junta de Andalucía. https://www.juntadeandalucia.es/organismos/fomentoarticulaciondelterritorioyvivienda/areas/vivienda-rehabilitacion/planes-instrumentos/paginas/vivienda-bcca.html. Accessed 19 Feb 2023
Lopes R. Silva D, Rivero-Camacho C, Rusu D, Marrero M (2022) Methodology for improving the sustainability of industrial buildings via matrix of combinations water and carbon footprint assessment. Sustainability 14:15297. https://doi.org/10.3390/su142215297
Marrero M, Rivero-Camacho C, Martínez-Rocamora A et al (2024) Holistic assessment of the economic, environmental, and social impact of building construction. Application to housing construction in Andalusia. J Clean Prod 434:140170. https://doi.org/10.1016/j.jclepro.2023.140170
Martínez-Rocamora A, Solís-Guzmán J, Marrero M (2016) Toward the ecological footprint of the use and maintenance phase of buildings: utility consumption and cleaning tasks. Ecol Indic 69. https://doi.org/10.1016/j.ecolind.2016.04.007
Meng J, Chen GQ, Shao L et al (2014) Virtual water accounting for building: case study for E-town, Beijing. J Clean Prod 68:7–15. https://doi.org/10.1016/j.jclepro.2013.12.045
Miguel CMG (2011) Real Decreto 1492/2011, de 24 de octubre, por el que se aprueba el Reglamento de valoraciones de la Ley de Suelo (BOE núm. 270, de 9 de noviembre). Actualidad Jurídica Ambiental
Ministerio de la Presidencia del gobierno de España (2017) Ley 9/2017, de 8 de noviembre, de Contratos del Sector Público, por la que se transponen al ordenamiento jurídico español las Directivas del Parlamento Europeo y del Consejo 2014/23/UE y 2014/24/UE, de 26 de febrero de 2014
Omniclass (2012) Omniclass: A strategy for classifying the built environment – Table 13: Spaces by function
Pastor Sánchez J (2015) PROCEDIMIENTO DE ESTIMACIÓN DEL COSTE DEL CICLO DE VIDA DE UN SISTEMA DE ARMAS EN España
PRé Sustainability (2016) SimaPro 8
PRECIOCENTRO (2023) Generador de precios descompuestos (Decomposed price generator), Guadalajara. In: Precio Centro de la Construcción.. https://preciocentro.com/. Accessed 12 Mar 2023
REE (2014) El sistema eléctrico español / The Spanish electric system
Rivero-Camacho C, Marrero M (2022) Water footprint of the life cycle of buildings: case study in Andalusia, Spain. In: Ren J (ed) Advances of footprint family for sustainable energy and industrial systems. Green energy and technology. Springer, Cham, pp 135–165. https://doi.org/10.1007/978-3-030-76441-8_7
Rivero-Camacho C, Martín-del-Río JJ, Marrero-Meléndez M (2023) Evolution of the life cycle of residential buildings in Andalusia: economic and environmental evaluation of their direct and indirect impacts. Sustain Cities Soc 93:104507. https://doi.org/10.1016/j.scs.2023.104507
Royal Decree 105/2008 (2008) National Decree 105/2008, February 1, which regulates the production and Management of Construction and Demolition Waste. Ministry of Presidency, Madrid
Ruiz-Pérez MR, Alba-Rodríguez MD, Marrero M (2022) Evaluation of water footprint of urban renewal projects. Case study in Seville, Andalusia. Water Res 221:118715. https://doi.org/10.1016/j.watres.2022.118715

Sherif YS, Kolarik WJ (1981) Life cycle costing: concept and practice. Omega (Westport) 9:287–296. https://doi.org/10.1016/0305-0483(81)90035-9
Solís-Guzmán J, Marrero M, Ramírez-de-Arellano A (2013) Methodology for determining the ecological footprint of the construction of residential buildings in Andalusia (Spain). Ecol Indic 25:239–249. https://doi.org/10.1016/j.ecolind.2012.10.008
Solís-Guzmán J, Rivero-Camacho C, Alba-Rodríguez D, Martínez-Rocamora A (2018) Carbon footprint estimation tool for residential buildings for non-specialized users: OERCO2 project. Sustainability (Switzerland) 10. https://doi.org/10.3390/su10051359
SpainGBC LEED (2017) LEED. In: http://www.spaingbc.org/web/leed-4.php
TGO – Sustainability of construction works (2021) EN 15643:2021 Sustainability of construction works - Framework for assessment of buildings and civil engineering works
UNE-EN 15978:2012 (2012) Sustainability of construction works. Assessment of environmental performance of buildings. Calculation Method
United Nations (2012) UN General Assembly Resolution on The Future we want (adopted on 27 July 2012)
Vivienda [C]. https://jmrarquitectos.com/proyectos/vivienda-en-espartinas/. Accessed 20 May 2024
Woodward DG (1997) Life cycle costing—theory, information acquisition and application. Int J Proj Manag 15:335–344. https://doi.org/10.1016/S0263-7863(96)00089-0

Sustainable Brine Management and Carbon Dioxide Capture: Minimizing Water Footprint, Addressing Environmental Challenges, and Promoting Resource Recovery

Jawad Mustafa, Muhammad Ahmad, Muhammad Imran, and Muhammad Wakil Shahzad

1 Introduction

Water, which is crucial for all living beings, is a highly valuable resource. Fortunately, it is also abundant on our planet, covering approximately 71% of the Earth's surface (How Much Water is There on Earth? | U.S. Geological Survey 2019). Water resources can be broadly classified as natural and non-conventional. Natural water resources are further categorized into saline waters and freshwaters based on their salinity levels. Freshwater has low salinity and is acknowledged as a renewable water resource owing to its natural water cycle. Surface freshwater, such as that found in fresh lakes and rivers, has low salinity, making it suitable for drinking. However, despite its common occurrence, surface freshwater accounts for only 0.27% of the total freshwater available (Gleick et al. 2006). Beneath the Earth's crust lies fresh groundwater, which distinguishes itself from surface freshwater through its elevated salinity levels (Mickley, and Associates, Environmental Resources Team (U.S.) 2018). It represents approximately 0.76% of the total

J. Mustafa
Mechanical & Construction Engineering Department, Northumbria University, Newcastle, UK

M. Ahmad · M. W. Shahzad (✉)
Mechanical & Construction Engineering Department, Northumbria University, Newcastle, UK

EcoTechX Limited (Company number 15011377) - A Spinout from Northumbria University, Newcastle, UK
e-mail: muhammad.w.shahzad@northumbria.ac.uk

M. Imran
Department of Mechanical, Biomedical and Design Engineering, Aston University, Birmingham, UK
e-mail: m.imran12@aston.ac.uk

S. S. Muthu (ed.), *Sustainability and Water Footprint*, Environmental Footprints and Eco-design of Products and Processes, https://doi.org/10.1007/978-3-031-70810-7_6

water on Earth and about 30% of the total freshwater (Gleick et al. 2013). Unfortunately, a considerable portion of this freshwater is confined within glaciers and permanent snow caps, making it unattainable for immediate use. Conversely, saline water, prevalent in oceans, seas, and saltwater lakes, encompasses the predominant share of water reserves on our planet, constituting approximately 97.5% of the overall water supply (Gleick et al. 2013). Considering the constrained accessibility of freshwater, the process of desalination has emerged as a prospective remedy for water provision in nations facing insufficient or constrained access to freshwater reservoirs. Desalination entails the treatment of saline water to procure freshwater, presenting a promising solution for addressing water scarcity challenges (Elsaid et al. 2020).

The process of desalination, which involves treating saline water sources like brackish or seawater to obtain potable water, has made significant advancements both economically and technically since its evolution in the 1950s. These advancements have contributed to a remarkable increase in global desalination capacity. In 2005, the global desalination capacity was approximately 35 million m^3 per day. However, due to ongoing improvements in desalination technologies and processes, this capacity has expanded to reach 95 million m^3 per day. This substantial growth reflects the growing demand for freshwater resources in various regions around the world. Of the total desalinated water produced, a significant portion, about 62.2%, is utilized for municipal purposes, catering to the needs of growing urban populations. This includes drinking water supply for residential areas, as well as water for public amenities such as schools, hospitals, and parks. The remaining 30.2% of desalinated water is used for industrial purposes, supporting sectors such as power generation, agriculture, and manufacturing (Gleick et al. 2013; Jones et al. 2019). When considering the distribution of desalination capacity, the Middle East and North Africa (MENA) region, which includes the Gulf Cooperation Council (GCC) countries, account for about 18.4% of the capacity. The East Asia and Pacific regions hold the largest share at 47.5%, and North America follows with 11.9% of the capacity (Elsaid et al. 2020; Jones et al. 2019).

In this chapter, we discuss the simultaneous process of brine management and CO_2 capture based on electrodialysis in detail. We begin with an overview of conventional brine production methods and their impacts on water footprint, followed by a brief mention of the elemental composition of seawater brine. The general principles and methods of electrodialysis are then discussed. Subsequently, we review various studies that focus on the combined management of brine and CO_2 capture using electrodialysis.

The chapter then delves into the pretreatment of brine using selective electrodialysis (SED). We provide detailed information on the use of treated brine in a bipolar membrane electrodialysis (BMED) process, wherein CO_2 is injected to form a range of value-added products. Given the limitations of BMED, we also explore an alternative technique: batch circulation electrodialysis. This method can generate an alkaline environment suitable for CO_2 capture and serves as a viable alternative to BMED. Finally, we present conclusions and future outlooks, offering guidance for further exploration in this area.

1.1 Desalination Technologies

Numerous technologies are available for desalinating both seawater and brackish water sources. While many of these technologies are commercially established, there are also emerging options like membrane distillation, which are either approaching commercialization or in the advanced stages of development, but they are not specifically assessed in this analysis. When determining the appropriate desalination technology, several factors come into play, including the quality of the source water, energy demands, and, of utmost significance, the perceived value of the treated water. The osmotic pressure fluctuates according to the salinity levels present in the water. Seawater, exhibiting a salinity of 35,000 mg/L, possesses an osmotic pressure of 2800 kPa, while brackish water, with a salinity of 1600 mg/L, exhibits a relatively lower osmotic pressure of 140 kPa (Xie et al. 2018). This difference in osmotic pressure has implications for technologies like reverse osmosis (RO). In the process of treating seawater, it becomes essential to apply significantly greater pressure to counteract the osmotic transfer of water across the semi-permeable membrane. Consequently, the energy demands for desalinating seawater are considerably more substantial in comparison to the treatment of brackish water. While grid energy serves as the primary energy source for desalination, there is also the possibility of employing alternative energy sources. These alternatives encompass solar, wind, wave, and thermal sources. By harnessing these renewable energy sources, it becomes feasible to acquire the requisite power for driving the desalination process (Forstmeier et al. 2007; Ettouney and Rizzuti 2007).

Solar desalination technology already exists and has been successfully applied in South Australia to provide water, energy, and heat for high-value crop production. Additionally, several demonstration plants employing photovoltaic-powered RO have been developed (Papapetrou et al. 2007). The direct conversion of wind power to mechanical energy is also being studied in the Netherlands. Wave energy is another potential source of direct mechanical energy, and a hydrostatic pressure-based system has been tested as well (Al-Kharabsheh 2006).

Water quality and water footprint play a crucial role in selecting appropriate desalination technologies, with salt concentration being a particularly important factor. The choice of process is also influenced by fouling and scaling, which depend on the composition of the source water. Technologies like RO commonly face practical challenges such as membrane deterioration, especially due to biofouling and scaling, which can be more pronounced during intermittent operation (Burn et al. 2015).

1.2 Brine Production and Water Footprint

One significant challenge linked with desalination is brine production, a concentrated solution that contains the salts and impurities removed from the source water during the desalination process. Nearly all desalination technologies generate a substantial amount of brine, often in quantities similar to or even greater than the volume of freshwater produced. The production of brine is an inevitable consequence of desalination because the salts and minerals present in the feedwater need to be separated and concentrated to obtain freshwater. The concentration of salts in brine can be several times higher than that of the source water, making it a highly concentrated and saline solution. Handling and disposing of brine are critical aspects of desalination plant operations. Concentrated brine has the potential to harm marine ecosystems if discharged into the ocean without proper treatment and dilution. The high salinity levels and increased concentrations of chemicals and metals in the brine can adversely affect marine life and the overall ecological balance. Moreover, this discharge significantly amplifies the water footprint of desalination operations, intensifying their overall environmental impact. Picture this: the discharged brine infiltrates the nearby water sources like rivers, lakes, or underground aquifers, contaminating them with its high salinity. This pollution renders the water less suitable for crucial needs like drinking, agriculture, or sustaining aquatic ecosystems (Missimer and Maliva 2018). The sheer volume of brine produced relative to the fresh water output significantly contributes to the water footprint of desalination. For every unit of fresh water generated, a substantial amount of brine is produced, often at a ratio of 1.5 to 2 times. Managing and disposing of this brine requires additional resources and infrastructure, thus amplifying the overall water footprint of the desalination process.

A typical brine resulting from the process of RO exhibits a concentration of total dissolved solids (TDSs) of approximately 70,000 mg L^{-1}. This brine comprises of various chemical residues originating from pretreatment procedures, along with sodium chloride (Kaplan et al. 2017; Al-Sarkal and Arafat 2013). During the pretreatment stage, chemicals such as $FeCl_3$, NaOCl, $AlCl_3$, HCl, H_2SO_4, and $NaHSO_3$ are commonly utilized to assist in water treatment. These chemicals serve multiple purposes, including reducing algal growth, inhibiting scaling, reducing corrosion, adjusting pH levels, and water chlorination (Burn et al. 2015).

1.3 Conventional Brine Disposal Methods

Within the desalination sector, multiple alternatives have been adopted to address the management of substantial volumes of reject brine produced during the desalination process. The selection of a disposal method commonly relies on factors such as the geographical placement of the desalination plant and the specific techniques employed in the process. Various possibilities exist for managing the substantial quantities of reject brine generated in the desalination process. These alternatives encompass discharging the brine into surface water bodies or wastewater treatment facilities, employing land disposal methods, implementing deep well injection, utilizing evaporation ponds, or resorting to thermal or mechanical evaporation techniques. The management of reject brine has become progressively more complex due to several factors. These factors include the expanding number and scale of desalination plants, which subsequently restricts the available options for disposal. Furthermore, the implementation of more stringent regulations concerning discharges adds further complexity to the disposal process, while an increased public awareness and concern regarding environmental matters amplifies the challenges. Additionally, the proliferation of desalination plants in semi-arid regions poses limitations on conventional methods of brine disposal. The consideration of cost is of utmost importance when determining the appropriate method for brine disposal, as estimates indicate that it can encompass a significant portion, ranging from 5% to 33%, of the overall cost of desalination. Several factors come into play when selecting the most suitable method for brine disposal. These factors include the volume and characteristics of the brine, composition of the concentrate, geographical or physical location of the discharge point, availability of appropriate receiving sites, permissibility of the chosen option according to regulations, capital and operational expenses, public acceptance, and the potential for future facility expansion. The expenses associated with brine disposal are contingent upon a range of factors, including the properties of the reject brine, the extent of treatment necessary prior to disposal, the chosen method of disposal, the volume of brine requiring disposal, and the specific characteristics of the disposal environment. Mickley (Mickley 2000, 2004) has compiled a comprehensive report that extensively examines various approaches to brine disposal. In the following sections, a concise summary will be presented regarding the principal alternatives for brine disposal, along with an outline of their main limitations and drawbacks.

Deep well injection has traditionally been utilized for the disposal of municipal, industrial, and liquid hazardous wastes (Saripalli et al. 2000). In recent times, deep well injection has emerged as a viable alternative for brine disposal, particularly for inland desalination plants, where surface water discharge is impractical or not economically feasible. Deep wells present a pragmatic and dependable approach for disposing of reject brine. Nonetheless, their viability is constrained in regions susceptible to seismic activity or where faults are present, as these geological circumstances can establish a direct hydraulic linkage between the receiving aquifer and a potable aquifer located above (Mickley 2014). Therefore, a thorough assessment of the geological conditions is necessary before drilling an injection well to determine appropriate porous aquifer reservoirs' depth and location (Berreteaga et al. 2013). The implementation of deep well injection also contributes to the negative impact on the water footprint. While it offers a solution for brine disposal, it can pose risks such as potential contamination of groundwater sources and soil degradation. Consequently, the use of deep well injection adds to the overall environmental burden associated with desalination, further amplifying its water footprint. Deep well injection generally incurs higher initial capital costs compared to surface water disposal due to the elimination of long brine transport pipelines. Although deep well injection can be considered as a viable solution for reject brine disposal, it is not without its limitations and drawbacks. These include the difficulty in identifying a suitable well site, the additional expenses associated with treating the reject brine before injection, the potential for corrosion and subsequent leakage in the well casing, and the risk of seismic activity causing damage to the well and potential contamination of groundwater (Berreteaga et al. 2013). Researchers have conducted thorough investigations into the performance, modelling aspects, and design considerations of deep well injection (Saripalli et al. 2000; Skehan and Kwiatkowski 2000; Rhee et al. 1993). These studies aim to address various aspects of deep well injection to ensure its effectiveness and minimize the potential risks associated with this disposal method.

Evaporation ponds have been recognized as a highly efficient and cost-effective approach for brine disposal in inland desalination plants, particularly in dry and arid regions akin to those prevalent in the Middle East and North Africa. In these specific regions, inland desalination plants are commonly located in areas that experience high temperatures, arid climatic conditions, and consequently, accelerated rates of evaporation. Ahmed et al. (2000) undertook a comprehensive analysis of existing literature, with a specific focus on the design considerations associated with evaporation ponds. Their study highlighted the critical importance of carefully selecting key design parameters, including surface area and pond depth, to ensure optimal performance and efficiency (Ahmed et al. 2000). In another investigation conducted by Ahmed et al. (2001), the application of evaporation ponds in Arabian Gulf countries, specifically Oman and the United Arab Emirates, was examined. The study revealed that newer desalination plants in the region employ lined evaporation ponds, whereas older plants still rely on unlined disposal pits for brine management (Ahmed et al. 2001). The primary environmental issue related with the disposal of brine through evaporation ponds is the potential risk of pond leakage, which has the potential to result in groundwater contamination in the surrounding vicinity. To mitigate this concern, contemporary evaporation ponds are commonly equipped with liners made of polymeric materials. The purpose of these liners is to prevent the leakage and seepage of contaminants into the adjacent groundwater, ensuring a higher level of environmental protection. The rate of evaporation plays a pivotal role in determining the efficiency of evaporation ponds, and it is significantly influenced by prevailing weather conditions. Efforts have been made to augment evaporation rates using techniques such as wind-aided intensified evaporation. These methods aim to enhance the evaporation process by leveraging the assistance of wind and other factors to increase the overall evaporation efficiency (Gilron et al. 2003), but the success of these methods has been limited. Wind-aided intensified evaporation proposes a 50% increase in evaporation rates in arid environments, although it remains dependent on favorable weather conditions. Enhancing the evaporation rate offers the potential to reduce the footprint of evaporation ponds, improve their effectiveness, and broaden their suitability across diverse regions globally. Nevertheless, despite the advantages of elevated temperatures and accelerated water evaporation, evaporation ponds still present drawbacks such as the requirement for extensive land area and the potential for contaminants to disperse into the soil and groundwater. Evaporation ponds rely on natural evaporation to concentrate the brine and reduce its volume. However, this process does not require additional water; instead, it utilizes the brine itself. In arid or water-stressed regions, where desalination plants are often located, the use of evaporation ponds can further strain already limited water resources indirectly. This strain arises because the water evaporated from the brine in the ponds is essentially lost from the local water cycle, as it does not return to the available water supply. Consequently, while evaporation ponds themselves do not directly consume water from external sources, they contribute to overall water scarcity by removing water from the local hydrological system. This further increases the water footprint of the desalination process, particularly in regions where water resources are already under stress.

1.4 Elemental Composition of Seawater and Brine

Table 1 provides insights into the concentrations of various elements in both seawater and brine. Sodium is the most abundant element, followed by magnesium, calcium, potassium, and other trace elements. The dominant anions in seawater are chloride, with small concentrations of sulfate and nitrate to maintain neutrality. Seawater is characterized by high levels of sodium, with concentrations around 10,800 ppm. Brine, a highly concentrated saline solution, exhibits even higher concentrations of sodium, ranging from 22,100 to 27,521 ppm. Magnesium concentrations in seawater are approximately 1290 ppm, increasing slightly to 2450–2867 ppm in brine. Calcium concentrations in seawater are around 411 ppm, with a range of 669–960 ppm in brine. Potassium concentrations are approximately 392 ppm in seawater, while in brine, they range from 554 to 783 ppm. Given the dominance of sodium and chloride ions, any treatment process that involves the removal of these elements requires careful consideration. It is important to account for the high concentrations and potential environmental impact associated with their discharge or disposal. Additionally, the treatment method should address the overall salinity and potential impacts on water quality and ecosystem balance.

Table 1 Composition of seawater and brine

Element	Concentration in seawater (ppm)	Concentration in brine (ppm)
Na	10,800	22,100–27,521
Ca	411	669–960
Mg	1290	2450–2867
Br	67.3	8–120
K	392	554–783
B	4.45	4.45
Sr	8.1	14.5–15.4
Rb	0.12	0.194–0.2
Li	0.17	0.27–0.4

Adapted from Ihsanullah Ihsanullah et al. (2022)

2 Electrodialysis-Based Approach for the Management of Reject Brine

The management of reject brine by electrodialysis using captured CO_2 is a new method based on the concept of valorization. This approach involves transforming the reject brine into value-added products. By doing so, the waste generated by desalination plants, namely the reject brine, can be effectively managed while maximizing its utilization. Consequently, this approach offers industries an incentive to view the reject brine as a resource rather than a mere waste.

2.1 *Electrodialysis Principles and Methods*

Electrodialysis is a process that employs ion-exchange membranes and an externally applied potential difference to facilitate the movement of ions. This results in concentrated ion solutions on one side of the membrane and depleted solutions on the other side. In contrast, passive dialysis utilizes semi-permeable membranes to remove or reduce solute concentrations. It operates by allowing solutes to move from areas of high concentration to areas of low concentration, resulting in decreased concentrations in one region and increased concentrations in another. Passive dialysis is commonly used for recovering molecules from dilute solutions and separating them from other molecules present in the original solution. This is achieved by employing membranes that selectively permit the passage of specific molecules while blocking others. On the other hand, electrodialysis focuses on the transfer of charges across ion-exchange membranes. These membranes are designed to primarily transport charged species, allowing for the removal of charges from one side to the other.

Electrodialysis operates on the principle of utilizing an alternating arrangement of cation exchange membranes (CEMs) and anion exchange membranes (AEMs), as depicted in Fig. 1. The process involves the application of an electric potential gradient, which can be generated either by electrochemical reactions at the electrodes or by a potential when no reaction takes place. This potential gradient drives the migration of ions through the stack of membranes, with anions and cations being introduced into the system. In the electrodialysis process, multiple sets of ion-exchange membranes are positioned between the cathode and anode at each end, allowing for the continuous separation and movement of ions. The typical electrochemical reactions that frequently occur at the electrodes are as follows:

$$\text{Cathode}: 2H^{+} + 2e - \rightarrow H_2$$

$$2H_2O + 2e - \rightarrow 2OH- + \ H_2$$

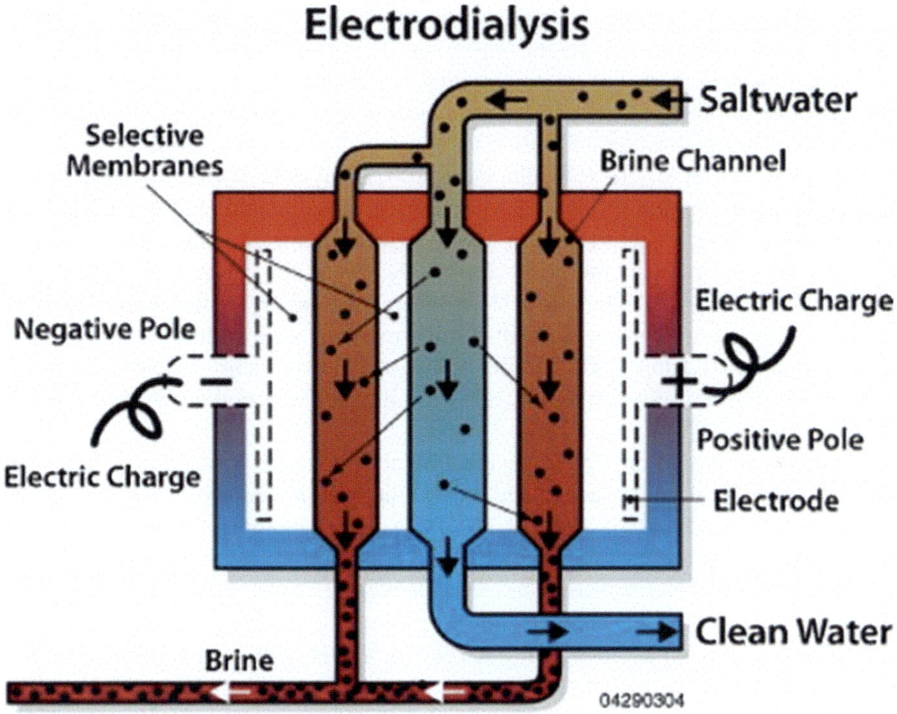

Fig. 1 Overview of the electrodialysis process. (Al-Karaghouli et al. 2009)

$$\text{Anode}: H_2O \rightarrow 2H^+ + 2e- + \ ½\,O_2$$

$$2OH- \rightarrow ½\,O_2 + 2H_2O + 2e-$$

AEMs are responsible for transporting anions towards the anode, while CEMs facilitate the movement of cations towards the cathode. Consequently, AEMs exclusively conduct anions, while CEMs exclusively conduct cations, aiming to achieve 100% current transfer within the electrochemical cell. However, in practice, a small fraction of the current carried by an AEM consists of cations (instead of anions), and for a CEM, a portion of the current involves anions (rather than cations) moving in the opposite direction of the intended operation. Nevertheless, membranes with high efficiency in current transfer may exhibit increased resistance, which can impact other important membrane properties. To enable the flow of solution between the AEM and CEM, a spacer is utilized. This spacer not only provides physical separation between the membranes but also induces turbulence in the solution, facilitating mass transfer. As a result of the removal of ionic species from the feed streams, the electrochemical cell is divided into alternating compartments, each containing a dilute stream and a concentrated stream. This partitioning allows for the effective separation and concentration of the targeted ionic species.

Electrodialysis offers a wide range of possibilities and can be applied in various configurations to achieve different objectives. This versatile technology allows for adjustment of the ion-exchange membrane patterns and quantities. Numerous applications of electrodialysis cells have been successfully demonstrated, including the treatment of brackish water and saline water, as well as the production of alkali and acid (Zabolotskii et al. 2014; Walha et al. 2007; Ghyselbrecht et al. 2013). It has proven to be an economically viable option for desalination. Electrodialysis has shown remarkable effectiveness in the removal of specific ions. For instance, reported removal rates include 65% for HCO_3^-, 91% for hardness, 72% for Cl^-, and 89% for SO_4^{2-} (Allen 2008). Moreover, electrodialysis has been found to consume less energy compared to alternative methods such as ion exchange using resin (Allen 2008; Turek 2004; Greiter et al. 2004). This energy efficiency contributes to the overall attractiveness and viability of electrodialysis for various applications.

Bipolar membranes utilized in the electrodialysis process have been successfully demonstrated to produce acids and bases through a water-splitting procedure (Ghyselbrecht et al. 2013; Jiang et al. 2014). By combining the produced base with CO_2, it becomes feasible to sequester CO_2 in the form of a mineral. Additionally, electrodialysis has proven effective in recovering caustic solvents employed for flue gas scrubbing in CO_2 capture processes (Iizuka et al. 2012). In this technique, CO_2 is absorbed in an alkaline solution, such as NaOH, within an absorption tower, resulting in the formation of a carbonate solution. The carbonate solution is then treated in an electrodialysis system, which allows for the recovery of gaseous CO_2 and regeneration of the alkaline solution. The regenerated solution can be reused for CO_2 capture within the absorption tower, making this approach suitable for CO_2 scrubbing utilizing solvents that require regeneration for CO_2 removal.

While electrodialysis offers various applications, it does have a fundamental limitation in achieving extremely low salinity levels, as discussed earlier. The presence of charged ions in the solution is essential for electrodialysis to occur since the process relies on an electrical potential to transport ions across membranes. As the concentration of ions in the solution decreases, the available current diminishes. Furthermore, as the ion concentration approaches zero, the required electrical potential increases to infinity. Consequently, electrodialysis cells are unable to produce pure water, but rather yield depleted solutions that still contain ions. However, by utilizing electrodialysis cells, it is possible to achieve NaCl concentrations below 2000 ppm.

2.2 *Electrodialysis to Achieve High Conversion Rates*

Electrodialysis, through the application of electrical potential, enhances the conversion rate of brine into valuable products. In a particular study (Park et al. 2015), CO_2 was transformed into various value-added products using sodium chloride (NaCl). An electrodialysis device operating at ambient temperature, low voltage, and

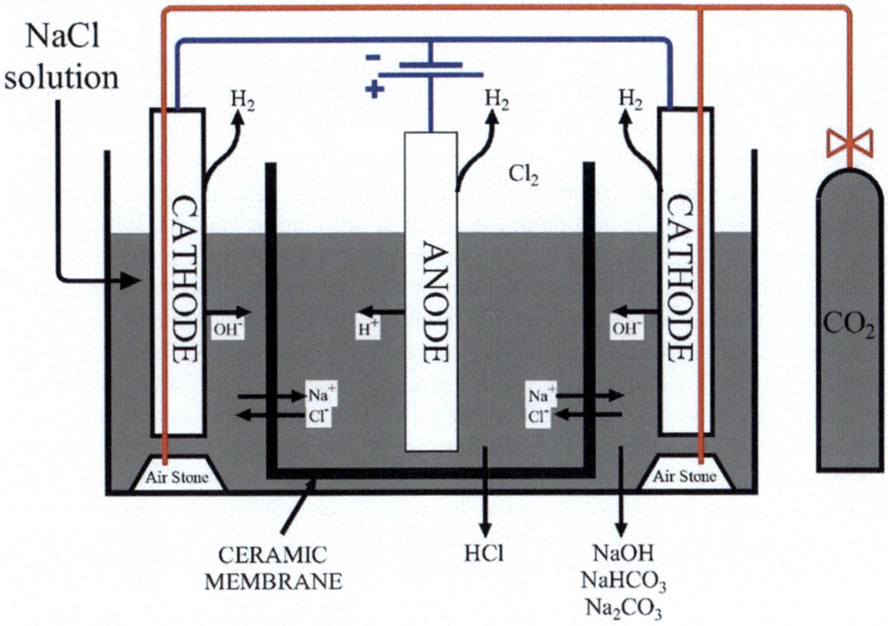

Fig. 2 Mineralization of CO_2 using electrodialysis. (Park et al. 2015)

atmospheric pressure was employed. Initially, NaCl was used to produce sodium hydroxide (NaOH) via electrodialysis. Subsequently, NaOH reacted with CO_2 to generate sodium bicarbonate ($NaHCO_3$). The overall process involves two primary reactions, resulting in the formation of sodium bicarbonate ($NaHCO_3$) and sodium carbonate (Na_2CO_3). To facilitate the reactions, a ceramic membrane divided the electrolytes into cathodes and anodes, as illustrated in Fig. 2. The use of a high-pressure air stone generated sufficient CO_2 bubbles, effectively increasing the contact area between the reactants (Park et al. 2015). The reactions occurring in this process are represented by Eqs. (1), (2), and (3).

Thermodynamic parameters provide evidence of the inherent spontaneity of these reactions.

$$2\mathrm{NaOH} + \mathrm{CO_2(g)} \leftrightarrow \mathrm{Na_2CO_3(aq)} + \mathrm{H_2O\ (l)} \quad \Delta H^0 = -169.8\ \mathrm{kJ/mol}, \quad \Delta G = -128.97\ \mathrm{kJ/mol} \tag{1}$$

$$\mathrm{Na_2CO_3(aq)} + \mathrm{H_2O\ (l)} + \mathrm{CO_2(g)} \leftrightarrow 2\mathrm{NaHCO_3(aq)} \quad \Delta H^0 = -129.1\ \mathrm{kJ/mol}, \quad \Delta G = -29.56\ \mathrm{kJ/mol} \tag{2}$$

The overall reaction is:

$$2\text{NaOH(aq)} + 2\text{CO}_2\text{(g)} \leftrightarrow 2\text{NaHCO}_3\text{(aq)} \quad (3)$$

Lee et al. conducted a study (Lee et al. 2018) investigating the utilization of the mineralization of CO_2 process, combining carbonation and brine electrodialysis, to facilitate the production of valuable chemical compounds, including hydrogen, sodium bicarbonate ($NaHCO_3$), and chlorine. The evaluation of the process's effectiveness involved the measurement of two key parameters: the rate of CO_2 reduction and the electric power consumption associated with brine electrodialysis. These metrics were used to assess the performance and efficiency of the process under investigation. To address the demanding operating conditions encountered in the brine electrodialysis system, a zero-gap membrane cell configuration was utilized. Additionally, the membrane design incorporated a thin layer of carboxylic acid to enhance its durability and resilience in the presence of strong acids, strong bases, and elevated temperatures. This approach aimed to ensure the system's stability and performance under challenging circumstances. The researchers reached the conclusion that the process demonstrated a remarkable purity level of over 97% for $NaHCO_3$. Furthermore, the CO_2 mineralization process resulted in the release of only 0.65 tons of CO_2 per ton of $NaHCO_3$ produced. This emission level is significantly lower compared to the 2.74 tons of CO_2 emissions associated with conventional processes used to produce the same quantity of $NaHCO_3$. In order to optimize the rate of the carbonation reaction and minimize both the size of the column and associated capital costs, the authors proposed the incorporation of carbonic anhydrase as a catalyst within the bicarbonation column. This suggestion aimed to enhance the efficiency of the process and improve cost-effectiveness by facilitating a more rapid reaction rate.

Researchers have identified the carbonation of sodium hydroxide (NaOH) as a viable option for reducing CO_2 emissions from flue gases. NaOH was generated through the chlor-alkali method using electrodialysis of sodium chloride (NaCl). By utilizing NaOH, a purity level of over 97% for $NaHCO_3$ was reached, and the rate of CO_2 absorption exceeded 95% (Shim et al. 2016; Jones and Yablonsky 2016). In another study (Lu et al. 2016), electrodialysis was found to significantly enhance the CO_2 capture efficiency. The process was studied in the presence and absence of electrodialysis. To raise the pH of the solution, fly ash was employed as an alkaline medium. Encouraging outcomes were observed when electrodialysis was incorporated into the process. The inclusion of electrodialysis, utilizing NaCl as an electrolyte, resulted in an 89% increase in CO_2 capture and a 34% improvement in fly ash dissolution. Due to the high concentration of Ca^{2+} in fly ash, the primary product obtained was calcium carbonate ($CaCO_3$). Moreover, the energy requirement for this process is anticipated to be lower than that of most combustion processes. Additionally, the utilization of hydrogen gas, which is already produced within the process, has the potential to further reduce the energy requirement.

Silicate minerals are abundant in nature and possess the inherent capability to undergo reactions with CO_2. This characteristic makes them an appealing choice for capturing CO_2 by converting it into carbonates and bicarbonates (Kelemen and

Matter 2008; Stephens and Keith 2008). However, the mineralization process that involves silicate minerals is generally characterized by slow kinetics under normal ambient conditions. To accelerate the dissolution of silicate minerals and facilitate their reactions with CO_2, electrodialysis can be utilized as an effective technique. In a study conducted by Rau et al. (2013), the electrochemical capture of CO_2 utilizing silicate minerals was investigated. The study utilized Na_2SO_4 as the electrolyte for the process. The predominant product obtained was $NaHCO_3$, along with metal sulphate, silica, and hydrogen. The researchers provided a representation of the potential net reactions involving silicate minerals, specifically wollastonite and ultramafic, in the presence of CO_2 and Na_2SO_4, as represented by Eqs. 4 and 5. It is worth noting that the reject brine, which commonly contains NaCl as a major component, could be employed as an electrolyte in this process, resulting in the production of metal chloride instead of metal sulphate. This approach enables the simultaneous and efficient management of both pollutants (CO_2 and reject brine).

$$2CO_2 + H_2O + MgSiO_{3(s)} + Na_2SO_{4\,(aq)} \rightarrow H_{2(g)} + 0.5O_{2(g)} + MgSO_{4(s)} + 2NaHCO_{3(aq)} + SiO_{2(s)} \tag{4}$$

$$2CO_2 + H_2O + CaSiO_{3(s)} + Na_2SO_{4\,(aq)} \rightarrow H_{2(g)} + 0.5O_{2(g)} + CaSO_{4(s)} + 2NaHCO_{3(aq)} + SiO_{2(s)} \tag{5}$$

The industrial implementation of the electrolysis process requires flexibility and stability to handle various saline waste fractions. In recent years, researchers have focused on improving the design characteristics of the process to enhance its practical feasibility (Dara et al. 2017; Park et al. 2015). In a study by Dara et al. (2019), a novel electrochemical cell design was presented, incorporating a platinum catalyst to facilitate the conversion of CO_2 into carbonate/bicarbonate salts. The authors described the electrochemical cell as consisting of five compartments, which were separated by CEMs and AEMs. The researchers investigated the impact of gas composition on the performance of the cell. They observed a significant reduction in cell performance as the CO_2 percentage increased, suggesting that catalyst poisoning could be a primary factor contributing to this decline. The specific composition and characteristics of the resulting products were influenced by the catalytic reduction of CO_2, with the catalyst type playing a vital role in determining the final outcomes. The utilization of an electrochemical cell consisting of five compartments showcased advantageous attributes such as flexibility, stability, and enhanced control over the entire process.

Dara et al. conducted a comparative analysis between the flux of product formation for carbonic acid (CO_2 dissolved in water) and CO_2 (without conversion into carbonic acid) (Dara et al. 2017, 2019). The researchers discovered that the direct utilization of CO_2 resulted in a product formation flux that was 40 times higher than that of carbonic acid. The dissimilarity in the product formation flux can be attributed to the limited solubility of CO_2 in water. To overcome the restricted solubility of CO_2 in water, a gas-fed electrode was utilized in the cathode compartment. This electrode facilitated the conversion of the gaseous mixture consisting of CO_2 and oxygen, along with saline water, into valuable products. By employing this approach, the challenges associated with the low solubility of CO_2 in water were effectively addressed. Figure 3 illustrates the configuration of the electrochemical cell employed in the study. The cell consists of an anode and a cathode, which are constructed using a titanium mesh coated with platinum (Pt) and iridium (Ir), respectively. These electrodes facilitate the production of $NaHCO_3$ and HCl from

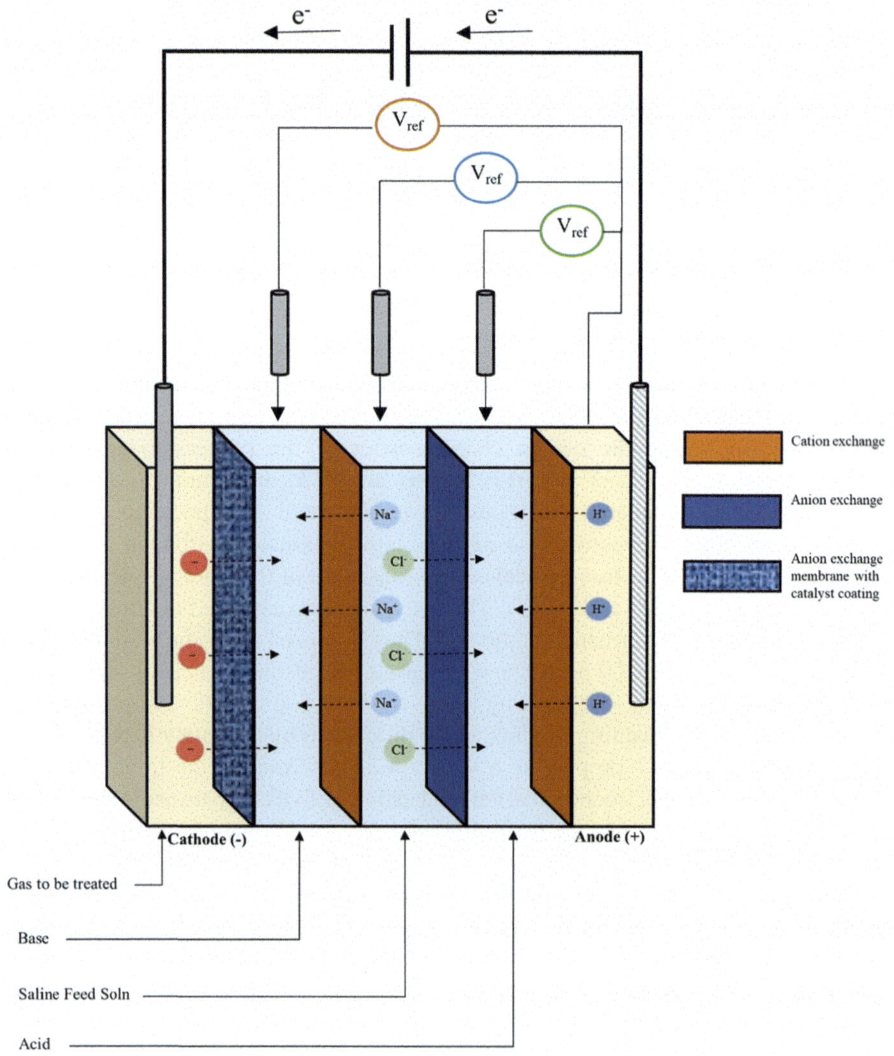

Fig. 3 Electrodialysis process using gas-fed electrode. (Dara et al. 2017, 2019; Mustafa et al. 2020)

the CO_2 and NaCl feed. Two AEMs and two CEMs were utilized in the cell. The AEM, coated with the catalyst, played a role in the conversion of CO_3^{2-}, HCO^{3-}, and OH^- species. To maintain the gap between the membranes, spacers were incorporated. Within the central compartments of the cell, three reference electrodes made of silver chloride were positioned. These reference electrodes were employed to assess potential losses occurring within the compartments. Despite achieving a high solubility for CO_2, the process still has certain limitations. One such limitation is the potential for catalyst poisoning caused by CO_2, which can lead to a decrease in the performance of the electrochemical cell.

In addition, the catalytic behavior varies depending on the current density, with most studies focusing on low current densities below 200 mA/cm^2. However, this limited focus on higher current densities may restrict its suitability for commercial-scale applications. Furthermore, before scaling up to industrial levels, it is necessary to investigate the purity of the $NaHCO_3$ and HCl produced. Practical feasibility is also constrained by high costs and membrane fouling. Moreover, the potential leakage of ions not only affects cell performance but can also lead to the formation of corrosive gases like chlorine, causing expensive electrode corrosion and increasing operational costs. Additionally, integrating this technology with existing setups poses a significant implementation challenge.

3 Unlocking the Potential: A Two-Stage electrodialysis Process for Sustainable Valorization of Industrial Brine with CO_2 and Water Footprint Minimization

Mustafa, Jawad et al. (2021, 2023, 2024a, b) proposed a comprehensive solution for treating raw reject brine solution using CO_2 in a two-stage process designed based on the zero-discharge principle, and this is further elaborated in Sect. 3. The authors presented two distinct electrodialysis designs: batch circulation and BMED. To pretreat the raw reject brine, the authors employed SED. A detailed description of the process developed by Mustafa, Jawad et al. is presented below (2021, 2023, 2024a, b). Commercial membranes obtained from PC-Cell GmbH were utilized in the process. The ion concentrations in the reject brine used in this study are as follows: Na^+ (17,000 ± 7 mg/L), Cl^- (31,000 ± 8 mg/L), SO_4^{2-}(5352 ± 4 mg/L), Ca^{2+} (586 ± 3 mg/L), Mg^2+ (2460 ± 6 mg/L), and K^+ (810 ± 3 mg/L).

The presence of multivalent ions, such as sulphate, magnesium, and calcium, can give rise to challenges such as membrane fouling and equipment damage. Therefore, it is essential to efficiently separate these ions from the reject brine. In the pursuit of establishing a circular economy, the simultaneous capture of carbon dioxide and effective management of the reject brine become of utmost importance. To tackle this challenge, Mustafa, Jawad et al. put forward a novel integrated approach that combines two membrane-based technologies: SED and BMED. This integration aims to optimize the valorization process, resulting in the generation of valuable

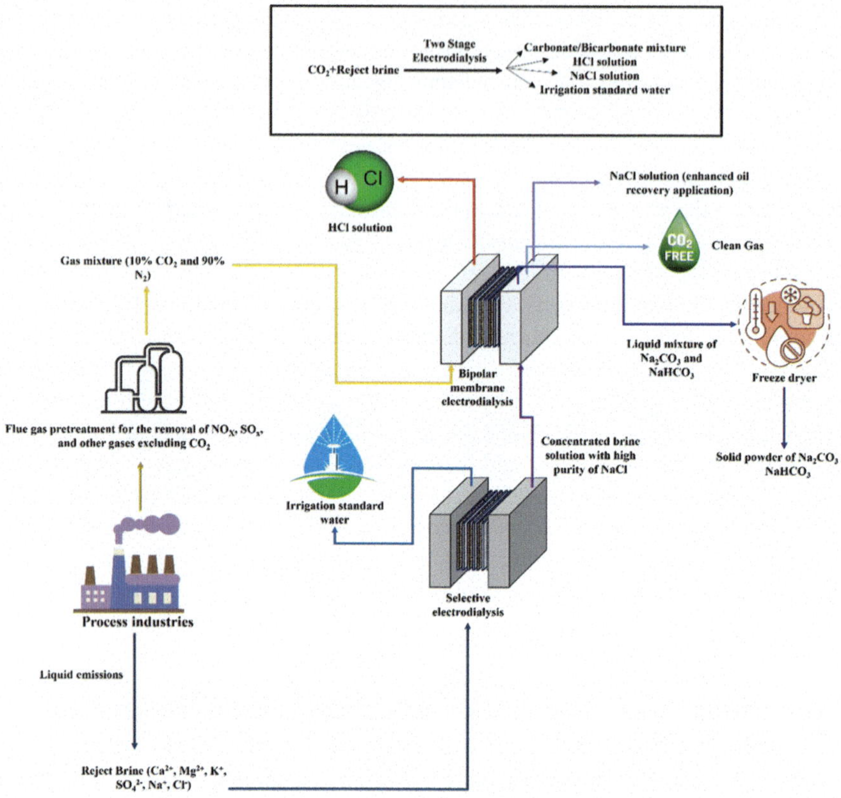

Fig. 4 Integration of SED with BMED for brine valorization. (Mustafa et al. 2024a)

products, including carbonate/bicarbonate salts, hydrochloric acid, irrigation-grade water suitable for plants capable of thriving in challenging conditions, and brine with a remarkable sodium chloride purity of 93%. Moreover, this process can result in significant reduction in the water footprint. Figures 4 and 5 illustrate the integration of SED with BMED for brine valorization. SED acts as a pretreatment unit for the brine, removing the multivalent fouling-prone species in liquid form from the brine. Subsequently, the NaCl stream is directed to the BMED process. To promote the smooth progress of the overall process and overcome initial resistance, a small concentration of various solutions was initially employed in all the chambers.

3.1 Description of the Process and Its Underlying Mechanism

Mustafa, Jawad et al. described the process of ion transfer across an ion exchange membrane (IEM) as a sequential series of five distinct stages (Mustafa et al. 2024a). To begin, ions that originate from the diluted compartment initiate their migration

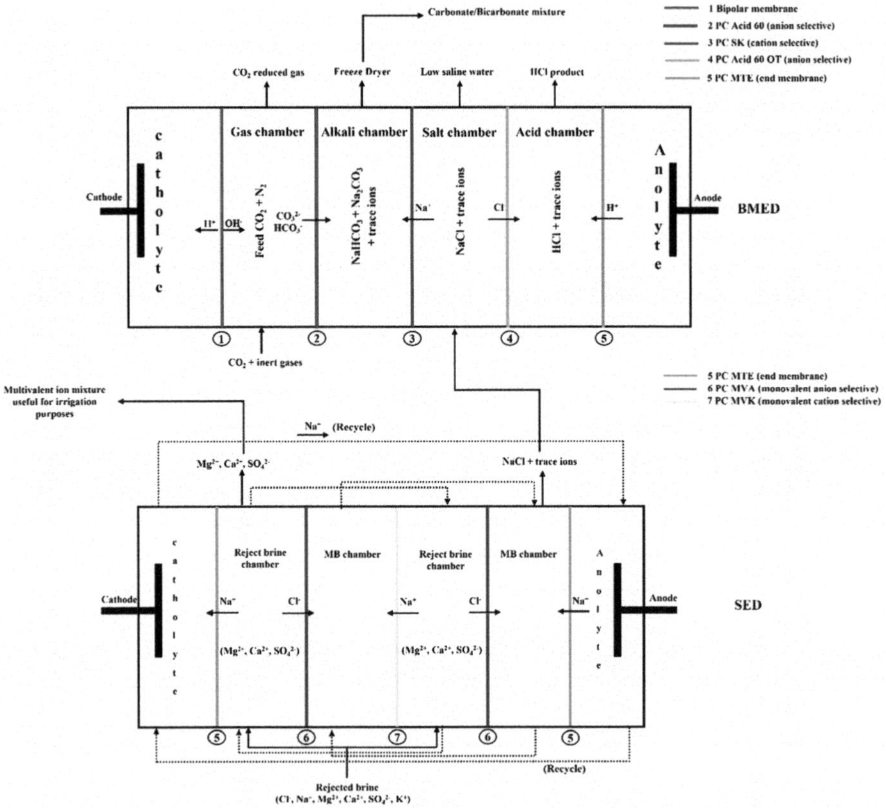

Fig. 5 Internal configuration of SED and BMED. (Mustafa et al. 2024a)

towards the surface of the membrane by traversing a boundary layer present within the solution. Following this, the ions proceed to partition into the membrane at its surface. The subsequent stage involves the movement of ions through the bulk of the membrane. Subsequently, in the fourth stage, the ions are released on the surface of the membrane. The selectivity of counter ions is predominantly governed by the process of the ionic species partitioning into the membrane and their subsequent migration through both the boundary layer and membrane matrix within the electrolyte solution.

The selectivity of ions during partitioning, especially in the presence of multiple competing counter ions, is influenced by various mechanisms. An example of such a mechanism is the electrostatic barrier effect, which emerges due to variations in the electrostatic interaction experienced by ions with the surface of the membrane. Another mechanism is related to the water affinity of ions, as characterized by their Gibbs hydration energy, and the hydrophobic properties of the membrane surface. These factors collectively contribute to a unique mechanism that influences the selectivity of ions during their permeation across the membrane. Additionally, steric hindrance plays a role in controlling the selectivity of counter-ions by restricting the partitioning of larger ions when a dense structure is present on the IEM. The mobility of ions is also a crucial factor that influences their selectivity within the membrane. Additionally, properties such as dimensions, hydrophobicity/hydrophilicity, and tortuosity of the ion pathway within the membrane have a notable influence on ion transport and, consequently, the selectivity of ions.

The selectivity in separating counter ions is affected by various operating conditions in electrodialysis, including cell pairs, current density, and flow rate. The movement of ions across an IEM positioned between electrodes is driven by the flow of faradaic current. As a consequence of this process, a thin layer is formed at the interface between the membrane and the solution, causing a depletion of ions in that region. The thickness of this boundary layer is variable and dependent on specific conditions, exerting an influence on the rate at which ions are transported towards the surface of the membrane.

When considering bipolar membranes, there is no consensus on the exact mechanism of water splitting in the transitional area. However, two plausible factors have been proposed to shed light on this process: (i) the second Wien effect and (ii) the catalytic effect of fixed ionogenic groups or other introduced chemicals. To decrease energy consumption during water splitting, modifications are made to the interface of the bipolar membrane, resulting in a catalytic effect. This approach is particularly beneficial, as it eliminates the production of gas or by-products, leading to a reduced voltage drop and enhanced energy efficiency. Moreover, the simultaneous operation of multiple bipolar membranes can be achieved with just one pair of electrodes, offering advantages such as lower initial investment, efficient space utilization, and simplified installation and operation. In the end, the same BMED stack enables the conversion of both inorganic and organic salts into their respective acids and bases.

In the process of BMED, there is a simultaneous treatment of the treated brine and capture of CO_2 from the gas mixture. The operation of the process followed a semi-continuous mode, where the gas mixture flowed continuously at different rates, while the flow of treated brine occurred intermittently or in a non-continuous manner. To maintain a constant circulation, each chamber, which included the salt, acid, alkali, and two electrolyte chambers, was linked to circulation cylinders. This arrangement ensured a continuous flow within the system. The purpose of this configuration was to minimize the thickness of the diffusion boundary layer and improve the transfer of ions. With the application of an electric field, bipolar membranes enabled the splitting of water molecules, resulting in the production of H^+ and OH^-. Within the gas chamber, a constant generation of OH^- took place, facilitating the CO_2 absorption from the gas mixture. This absorption led to the formation of HCO_3^- and CO_3^{2-}. In order to improve the absorption process, a standard membrane (PC Acid 60) was utilized. This membrane facilitated the entry of both HCO_3^- and CO_3^{2-} into the alkali chamber, enhancing the absorption of these ions. When the maximum amount of HCO_3^- and CO_3^{2-} are permitted to pass into the alkali chamber, it results in the highest absorption rate. Within the alkali chamber, the existence of elevated concentrations of HCO_3^- prompted a shift in the equilibrium towards the production of CO_3^{2-}. Na^+ from the salt chamber traversed the CEM and entered the alkali chamber, leading to the creation of a mixture comprising Na_2CO_3, $NaHCO_3$, and trace amounts of metallic carbonates, intermediates, and bicarbonates. This occurrence was attributable to the minimal leakage of low-concentration Ca^{2+} and Mg^{2+} into the alkali chamber.

$NaHCO_3$ and Na_2CO_3 equilibrium relied on the process conditions, and the purity of each compound could be controlled by adjusting these conditions accordingly. To overcome electrical resistance, a small concentration of Na_2CO_3 was initially introduced into the gas chamber. The H^+ produced through the dissociation of water by the bipolar membrane were utilized in the generation of HCl, along with other minor constituents. The predominant factors governing the movement of ions were diffusion and migration, while the influence of convection was minimal. The high concentration of Cl^- in the salt chamber caused a certain degree of diffusion of Cl^- into the alkali chamber. Therefore, it is essential to utilize a specialized type of CEM with a strong resistance to Cl^- in the salt chamber. This ensures effective prevention of Cl^- diffusion into the monovalent ion-concentrated brine (MB) chamber.

3.2 *Pretreatment of Brine Using SED*

The presence of Mg^{2+} concentrations measuring 2460 mg/L and Ca^{2+} concentrations of 586 mg/L indicates that the solution system has a propensity to create $MgCO_3$/Mg $(HCO_3)_2$ and $CaCO_3$/$Ca(HCO_3)_2$ compounds during BMED or batch circular mode electrodialysis (discussed in following sections). The low solubility of Ca^{2+}- and Mg^{2+}-based CO_3^{2-}/HCO_3^- results in their rapid precipitation, which can cause membrane fouling and blockage of pores. Hence, it is imperative to eliminate Mg^{2+} and Ca^{2+} from the reject brine solution. Under optimal conditions, the conductivity of the treated reject brine stream from the reject brine chamber was 30 mS/cm. The processed reject brine stream from the reject brine chamber comprises the following concentrations: Ca^{2+} at 500 mg/L, Mg^{2+} at 2305 mg/L, Na^+ at 4680 mg/L, Cl^- at 10400 mg/L, SO_4^{2-} at 3784 mg/L, and K^+ at 184 mg/L. The processed stream (reject brine chamber) from SED has a total concentration of 21,853 mg/L, making it suitable for irrigation purposes for certain plant species. Furthermore, SED increases the concentrations of Mg^{2+}, Ca^{2+}, SO_4^{2-}, and K^+ in the processed stream, which are essential nutrients for crop growth.

3.2.1 Parametric Influence on the Performance of SED

The current density in the ED process has a significant impact on the permselectivity of monovalent ions. Figure 6 depicts the variations in ion separation efficiencies over time at various current densities. Among the different ions, Mg^{2+} demonstrates the highest separation efficiency across all current densities, with Ca^{2+} and SO_4^{2-} following closely behind. At the beginning, the separation efficiency is relatively

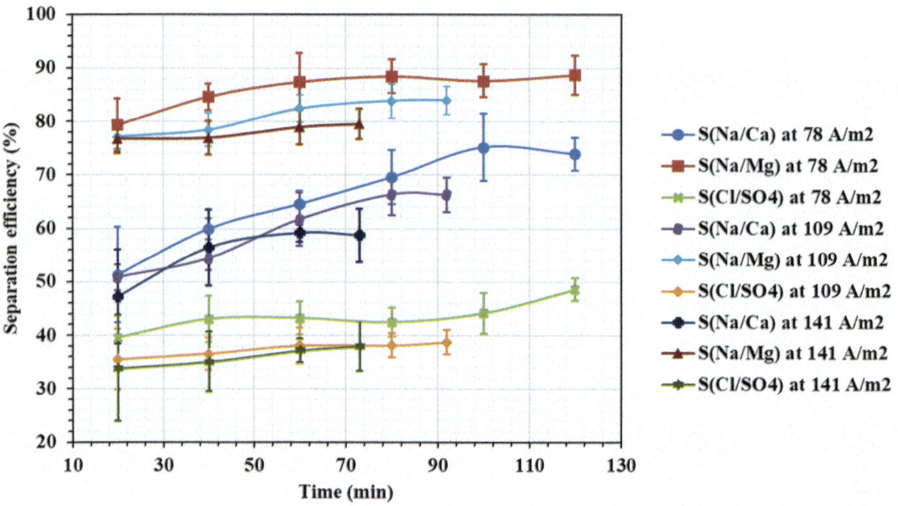

Fig. 6 Influence of current density on the separation efficiency of SED. (Mustafa et al. 2024a)

low owing to the limited concentration of Na^+ and Cl^- in the MB chamber. However, as more ions enter the chamber, the separation efficiency improves. After 120 min at a current density of 78 A/m^2, the separation efficiencies for S (Na/Mg), S(Na/Ca), and S(Cl/SO_4) were 89%, 74%, and 49% respectively. As the current density increases, the rate of ion transfer is accelerated; however, there is a corresponding decrease in separation efficiency. Processing times of 120, 84, and 80 min were achieved at current densities of 78, 109, and 141 A/m^2, respectively. At low current density, the transport of Ca^{2+}, Mg^{2+}, and $SO_4{}^{2-}$ from the reject brine chamber is impeded by Donnan exclusion. As the current density increases, the driving force on divalent ions intensifies, compelling them to enter the MB chamber under the influence of the electric field. This, in turn, leads to reduced separation efficiency. Monovalent selective membranes act as a barrier by creating electrical repulsion, thereby minimizing the penetration of divalent ions. However, at high current density, the electrostatic repulsion from the modified selective layer is partially compensated, resulting in a higher transfer rate of divalent ions.

Figure 7 illustrates the impact of the flow rate of the reject brine/MB chamber on separation efficiencies. The results demonstrate that altering the flow rate has a notable impact on the separation efficiencies of both monovalent and divalent ions. Generally, lower flow rates are associated with higher separation efficiencies. When the flow rate is set at 5 L/h, the transfer percentages of ions to the MB chamber are as follows: Ca^{2+} = 11%, K^+ = 75%, Mg^{2+} = 4%, Na^+ = 7.23%, Cl^- = 66.1%, and $SO_4{}^{2-}$ = 27%. On the other hand, when the flow rate is increased to 15 L/h, the transfer percentages of ions are: Ca^{2+} = 15%, K^+ = 77%, Mg^{2+} = 6%, Na^+ = 7.25%, Cl^- = 66.4%, and $SO_4{}^{2-}$ = 29%. Higher flow rates generally result in increased ion transfer. It is important to highlight that as the flow rate increases, the separation efficiency of monovalent ions decreases. This is primarily due to the

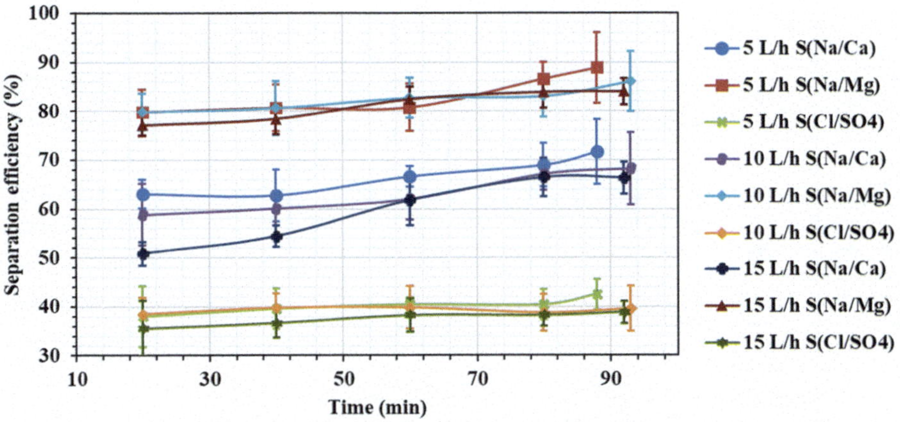

Fig. 7 Influence of flow rate of the separation efficiency of SED. (Mustafa et al. 2024a)

higher likelihood of divalent ions leaking into the MB chamber. Consequently, a higher flow rate compromises the selectivity of monovalent ions. The behavior can be explained by the fact that divalent ions, possessing higher charges, experience stronger electrostatic repulsion compared to monovalent ions at the membrane surface. With an increase in the flow rate, there is a corresponding rise in the external force exerted by divalent ions on the membrane surface. This causes a decrease in the electrostatic repulsion towards divalent ions and leads to their leakage into the concentrate chamber. As a result, the selectivity of monovalent ions, such as sodium, is reduced. At a flow rate of 5 L/h, the processing time required for the reject brine chamber to reach a conductivity of 30 mS/cm is 88 min. This processing time increases slightly to 93 min at a flow rate of 10 L/h and 92 min at a flow rate of 15 L/h. Furthermore, increasing the flow rate from 5 to 15 L/h leads to a decrease in separation efficiencies. Specifically, S(Na/Mg) decreases from 89% to 84%, S (Cl/SO_4) decreases from 43% to 39%, and S(Na/Ca) decreases from 72% to 66%.

The impact of cell pairs on separation efficiencies is illustrated in Fig. 8. The results indicate that higher separation efficiencies for monovalent ions are correlated with a lower number of cell pairs. When the number of cell pairs is increased, it also increases the membrane surface area, providing more opportunities for divalent ions to interact with the membrane. Consequently, there is an increased likelihood of divalent ion leakage. By increasing the number of cell pairs from 10 to 20, there is a decrease in the values of S(Na/Mg) from 87% to 84%, S(Na/Ca) from 73% to 66%, and S(Cl/SO_4) from 45% to 39%. This suggests a decline in the selectivity of monovalent ions as the number of cell pairs increases.

The impact of membrane type (selective/standard) on separation efficiencies is depicted in Fig. 9. Selective membranes were found to yield higher separation efficiencies compared to standard membranes. When selective membranes were utilized, the separation efficiencies were as follows: S(Na/Mg) = 84%, S (Na/Ca) = 66%, and S(Cl/SO_4) = 39%. In contrast, when standard membranes were used, the separation efficiencies were lower, with values of S(Na/Mg) = 54%,

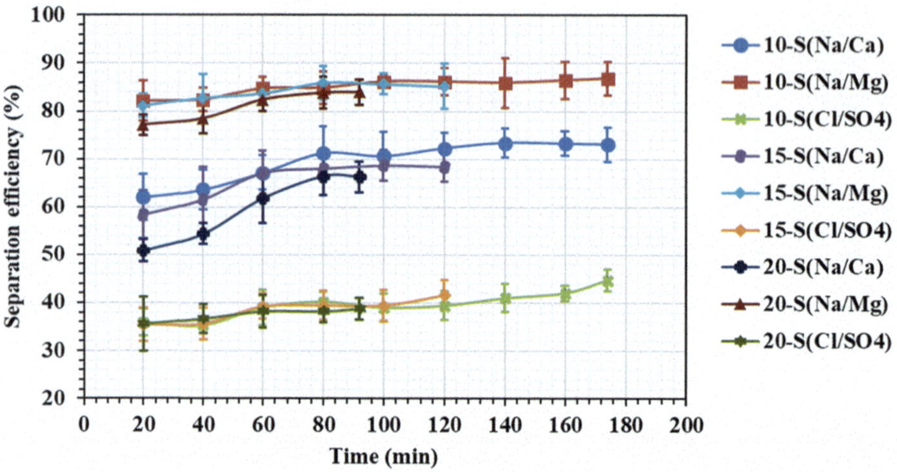

Fig. 8 Influence of cell pairs on the separation efficiency of SED. (Mustafa et al. 2024a)

S(Na/Ca) = 45%, and S(Cl/SO_4) = 36%. These results indicate that the use of selective membranes significantly improves the separation efficiencies of monovalent and divalent ions compared to standard membranes. Among the different separation efficiency parameters, the largest variations were observed in the case of S(Na/Mg), followed by S(Na/Ca) and S(Cl/SO4). This suggests that the selective membranes have a greater impact on the separation of sodium and magnesium ions, followed by the separation of sodium and calcium ions, and finally the separation of chloride and sulphate ions.

3.3 Bipolar Membrane Electrodialysis Process

Mustafa, Jawad et al. (2024a) discussed the process of simultaneously capturing CO_2 and treating the NaCl-dominant stream using BMED, as shown in Fig. 10. The NaCl-dominant stream utilized in this process was obtained from the reject brine, after the removal of multivalent ions through SED. The BMED stack comprised of four chambers, namely the gas chamber, salt chamber, alkali chamber, and acid chamber. It had a maximum of five cell pairs and two electrolyte cells. Each membrane had an active surface area of 0.0064 m^2. The distance between the membranes, which determines the flow channel width, was measured to be 0.00045 m. Additionally, the distance between the end membranes and the electrodes was found to be 0.001 m. Silicon/polyethylene sulfone wire mesh spacers with a void fraction of 59% were utilized to create different chambers. The electrodes in the BMED system consisted of a platinum-plated iridium anode and a titanium cathode. To maintain a continuous circulation of the concentrate and diluate solutions within the BMED stack, peristaltic pumps were employed in the external

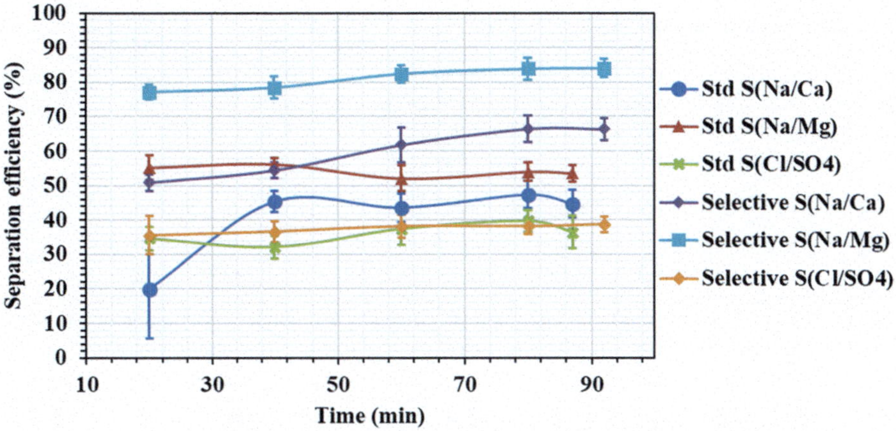

Fig. 9 Comparison of standard and selective membranes. (Mustafa et al. 2024a)

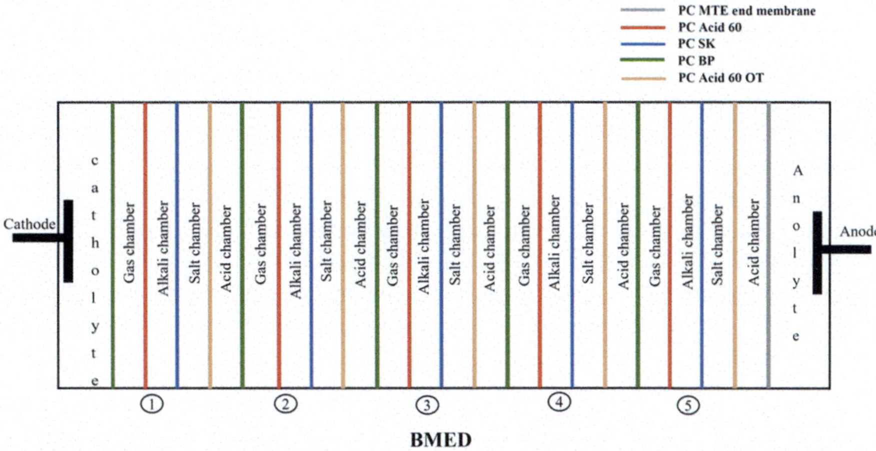

Fig. 10 Internal configuration of BMED. (Mustafa et al. 2024a)

circuits. To minimize the initial resistance, the BMED system was initially supplied with specific ion concentrations. The alkali chamber received an initial concentration of $NaHCO_3$, the acid chamber was provided with HCl, and the gas chamber was supplied with Na_2CO_3. The initial conductivities of the acid, alkali, gas, and salt chambers were measured to be 0.1 ± 0.04, 0.75 ± 0.09, 10 ± 0.2, and, 50 ± 0.5 mS/cm, respectively.

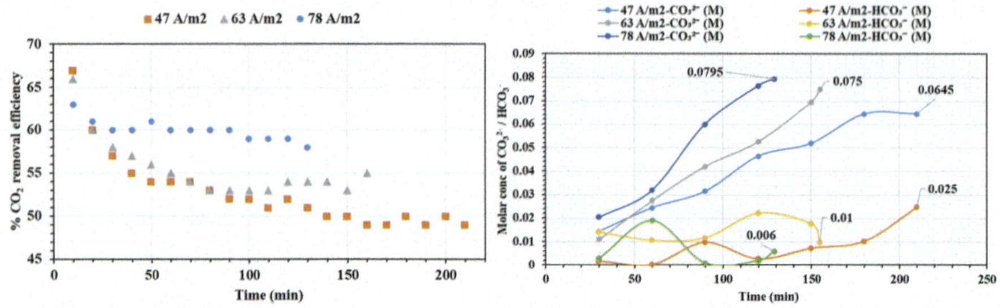

Fig. 11 Effect of current density on CO_2 removal efficiency and molar concentration of products. (Mustafa et al. 2024a)

3.3.1 Parametric Influence on the Performance of BMED

From Fig. 11, it is clear that the current density has a significant impact on the enhancement of CO_2 capture. As Na_2CO_3 is consumed, the CO_2 capture efficiency gradually decreases and finally stabilizes over time. The observed consistent efficiency in CO_2 capture suggests that bipolar membranes maintain a nearly constant rate of dissociation for water molecules. By increasing the current in a bipolar membrane, the dissociation of water into OH^- and H^+ is promoted, leading to improved CO_2 capture. When the salt chamber conductivity reached 35 mS/cm, CO_2 capture efficiencies of 49% at 47 A/m^2, 55% at 63 A/m^2, and 58% at 78 A/m^2 were attained. Figure 11 illustrates the variations in the HCO_3^-/CO_3^{2-} concentration within the liquid mixture present in an alkali chamber. At higher current densities, CO_2 absorption is enhanced due to the increased dissociation of water molecules in bipolar membranes. This leads to an elevation in the concentration of HCO_3^-, consequently shifting the equilibrium to favor a higher proportion of CO_3^{2-}. In particular, as the current density is raised from 47 to 78 A/m^2, there is an increase in the CO_3^{2-} concentration from 0.0645 M to 0.0795 M, whereas the HCO_3^- concentration decreases from 0.025 M to 0.006 M.

The results depicted in Fig. 12 demonstrate a decrease in the efficiency of CO_2 capture with an increase in the number of cell pairs. The enlargement of the membrane's active area is a consequence of employing additional cell pairs, which in turn reduces the conductivity of the salt chamber to 35 mS/cm more rapidly. Nevertheless, this augmentation in cell pairs may result in increased leakage and diffusion of ions through the membranes, ultimately leading to a gradual reduction in CO_2 capture efficiency. Specifically, at the 200-min mark, the respective CO_2 capture efficiencies for 3, 4, and 5 cell pairs are 55%, 50%, and 50%. Notably, the most significant reduction in CO_2 capture efficiency occurs when transitioning from 3 to 4 cell pairs. Furthermore, there is only a marginal disparity in CO_2 capture efficiency between cell pairs 4 and 5. Figure 12 presents a graphical representation of alterations in the concentration of HCO_3^-/CO_3^{2-} in an alkali chamber when varying the number of cell pairs. Upon completion of the process, the combined concentrations of HCO_3^- and CO_3^{2-} for 3, 4, and 5 cell pairs are measured as 0.114 M,

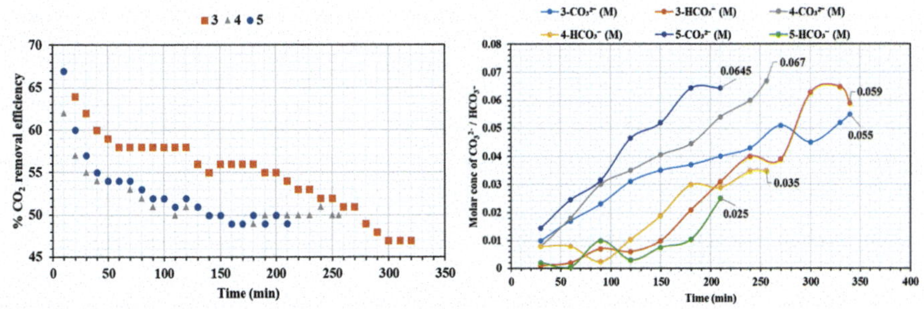

Fig. 12 Effect of cell pairs on CO_2 removal efficiency and molar concentration of products. (Mustafa et al. 2024a)

0.102 M, and 0.0895 M, respectively. The augmentation of cell pairs leads to a decrease in the HCO_3^-/CO_3^{2-} concentration due to irreversible processes and ion leakage. When high concentrations of HCO_3^- are present, the reaction tends to favor the forward direction, leading to an elevation in the CO_3^{2-} concentration. Conversely, in situations where HCO_3^- concentrations are low, the reaction tends to shift in the reverse direction, resulting in a decrease in the CO_3^{2-} concentration.

Based on the information provided in Fig. 13, there is a decrease in the efficiency of CO_2 capture with an increase in the flow rate of CO_2. The decline in current efficiency serves as clear evidence that a gas chamber operating at a current density of 47 A/m^2 lacks the capability to produce an adequate amount of OH^- ions necessary for effectively handling high CO_2 flow rates. Consistent with earlier trends, the CO_2 capture efficiency diminishes as Na_2CO_3 is consumed and finally stabilizes. Figure 13 exhibits the changes in HCO_3^-/CO_3^{2-} concentration in an alkali chamber in relation to the CO_2 flow rate. At the conclusion of the process, the combined concentrations of HCO_3^- and CO_3^{2-} for flow rates of 0.3, 0.5, and 0.7 L/min are measured as 0.0895 M, 0.105 M, and 0.1225 M, respectively. An increased CO_2 flow rate leads to higher concentrations of HCO_3^-/CO_3^{2-} in the alkali chamber as a result of the greater ionic flux of HCO_3^-/CO_3^{2-}. With an increase in the CO_2 flow rate from 0.3 to 0.7 L/min, there is an elevation in the HCO_3^- concentration from 0.025 M to 0.0825 M, while the CO_3^{2-} concentration decreases from 0.0645 M to 0.04 M. As a result, higher CO_2 flow rates cause a shift in the equilibrium towards the formation of HCO_3^-.

3.4 Batch Circulation Configuration: An Alternative Process for Bipolar membrane Systems in Brine Treatment, CO_2 Capture, Valorization, and Water Footprint Minimization

The batch circulation mode is an alternative configuration to the BMED process, where a bipolar membrane is employed for CO_2 absorption. Bipolar membranes,

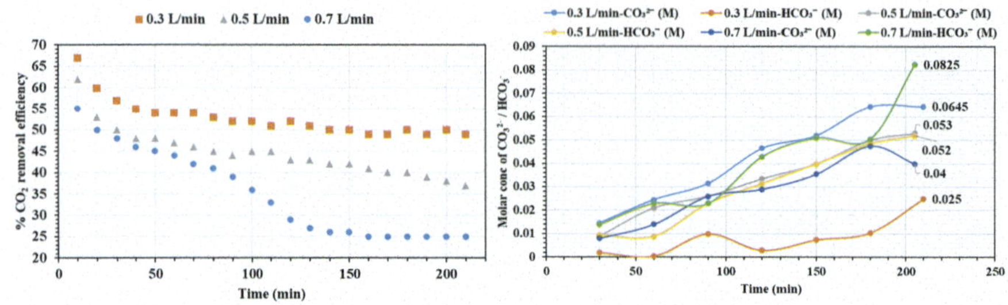

Fig. 13 Effect of CO_2 flow rate on CO_2 removal efficiency and molar concentration of products. (Mustafa et al. 2024a)

constructed by laminating a CEM and an AEM together, are two times as expensive as either a CEM or an AEM. Because of their drawbacks, including insufficient permselectivity, high levels of contamination, and short lifespan, bipolar membranes are infrequently utilized. Consequently, the batch circulation configuration has emerged as a novel process (Mustafa, Jawad et al. 2021, 2023), serving as an alternative to BMED. When subjected to an electric field, Na^+ migrate into the alkali chamber through the bipolar membrane, while Cl^- ions pass through the AEM into the acid chamber. Upon reaction with Na_2CO_3 and water, CO_2 is converted into $NaHCO_3$, resulting in the formation of a Na_2CO_3 solution within the gas chamber. These solutions are further transformed into Na^+, HCO_3^-, and CO_3^{2-}. Additionally, when Na^+ and OH^- combine in the cathodic cell, NaOH solution is generated. In the alkali chamber, $NaHCO_3/Na_2CO_3$ solutions are produced by the interaction of Na^+ from the salt chamber with HCO_3^- and CO_3^{2-}. During the absorption process, CO_3^{2-} and HCO_3^- are formed and transported through the AEM into the alkali chamber. The formulation of the $Na_2CO_3/NaHCO_3$ solution is contingent upon the particular process conditions. As Na_2CO_3 is utilized, the pH progressively decreases, leading to a reduction in CO_2 absorption. Eventually, the pH reaches a minimum value within the range of pH 7.5 to 8. To counteract this decline, the batch circulation mode is activated to boost CO_2 absorption. During the procedure, the batch circulation mode is initiated multiple times, depending upon the pH level of the gas chamber. The batch circulation entails the circulation of a 100-ml solution between the outer cylinders of the cathodic cell and the gas chamber. This process serves to increase the concentration of OH^-, thereby enhancing the absorption process.

Under optimal conditions, the liquid carbonate solution underwent freeze-drying to generate a solid mixture. The process of freeze-drying, in which an aqueous mixture of CO_3^{2-} and HCO_3^- is converted into a solid state, can alter the relative proportions of these components. This alteration is attributed to factors such as temperature, pressure, and the quantity of CO_2, which in turn impact the equilibrium of the system. It is noteworthy that even in an aqueous solution, modifying the operating conditions can lead to variations in the fraction of CO_3^{2-} and HCO_3^- within the mixture. In the anodic cell, cathodic cell, salt chamber, acid chamber, gas

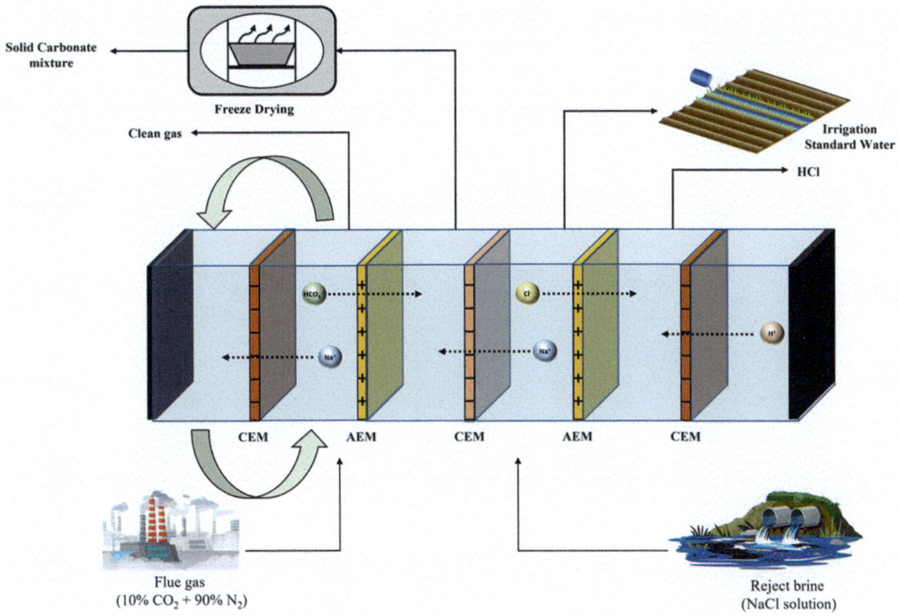

Fig. 14 Overview of the batch circulation configuration. (Mustafa et al. 2023)

chamber, and, alkali chamber the initial concentrations were as follows: 0.25 M H_2SO_4, 0.1 M NaOH, 0.33–1.17 M NaCl, 10 mM HCl, 0.1 M Na_2CO_3, and 10 mM $NaHCO_3$. Figures 14 and 15 depict a schematic representation of the overall process in the batch circulation configuration, providing an overview and specific details of the process, respectively.

3.4.1 Parametric Influence on the Performance of Batch Circulation Configuration

Figures 16 and 17 showcase the impacts of voltage/area, CO_2 flow rate, and salt concentration on various performance metrics such as waste-to-product ratio, CO_2 uptake (g/M NaCl), current efficiency for NaCl removal, and % salt removal. Figures 16 and 17 present surface contours that were created to illustrate the combined effects of two parameters while maintaining the third parameter constant. At the optimal conditions, a limiting current density of 194.276 A/m^2 was achieved.

The influence of various parameters on CO_2 uptake is depicted in Fig. 16a, b. Figure 16a demonstrates that increasing the voltage/area results in a rise in CO_2 uptake. This can be attributed to improved dissociation of gas within the chamber, leading to the subsequent generation of carbonic acid. With an increase in the voltage/area, there is a greater migration of HCO_3^- and CO_3^{2-} into the alkali chamber, leading to a more efficient capture of CO_2. However, the absorption

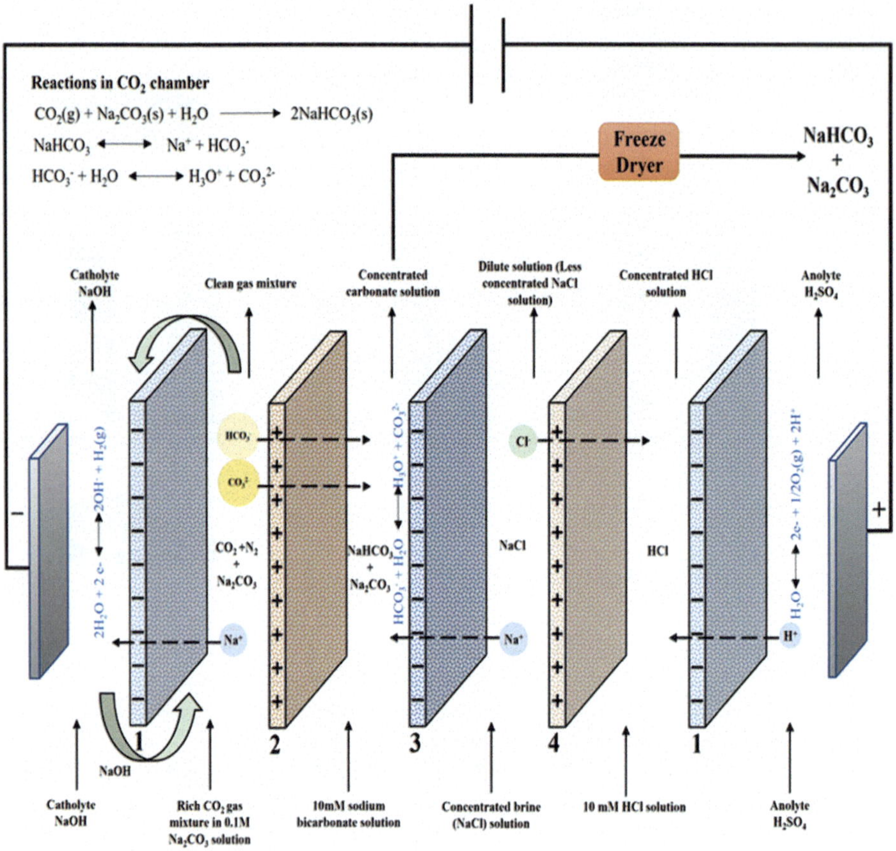

Fig. 15 In-depth detail of the batch circulation configuration. (Mustafa et al. 2023)

process experiences a gradual decline in the pH of the gas chamber, resulting in a deceleration of the absorption process. Through the utilization of batch circulation, the concentration of OH^- in the gas chamber is elevated, thus improving the absorption process. Moreover, within the cathodic cell, the concentration of OH^- increases with the voltage/area, thereby expediting the absorption process. Through the increase of voltage/area from 248 to 2876 V/m^2, while maintaining a CO_2 flow rate of 1 L/min, a rise in CO_2 uptake is observed from 6 to 13 g/M NaCl. This indicates that a twofold increase in CO_2 capture is achieved when the voltage/area is increased by a factor of 11. It is important to highlight that the absorption process does not show improvement beyond a voltage/area of 248 V/m^2. However, a notable increase in absorption is observed once the voltage/area reaches 781 V/m^2. Consequently, increasing the voltage leads to higher CO_2 uptake and a greater concentration of the carbonate mixture in the alkali chamber. Figure 16 also demonstrates that a higher CO_2 flow rate substantially enhances CO_2 uptake due to increased CO_2 concentration in the gas chamber. By increasing the CO_2 flow rate from

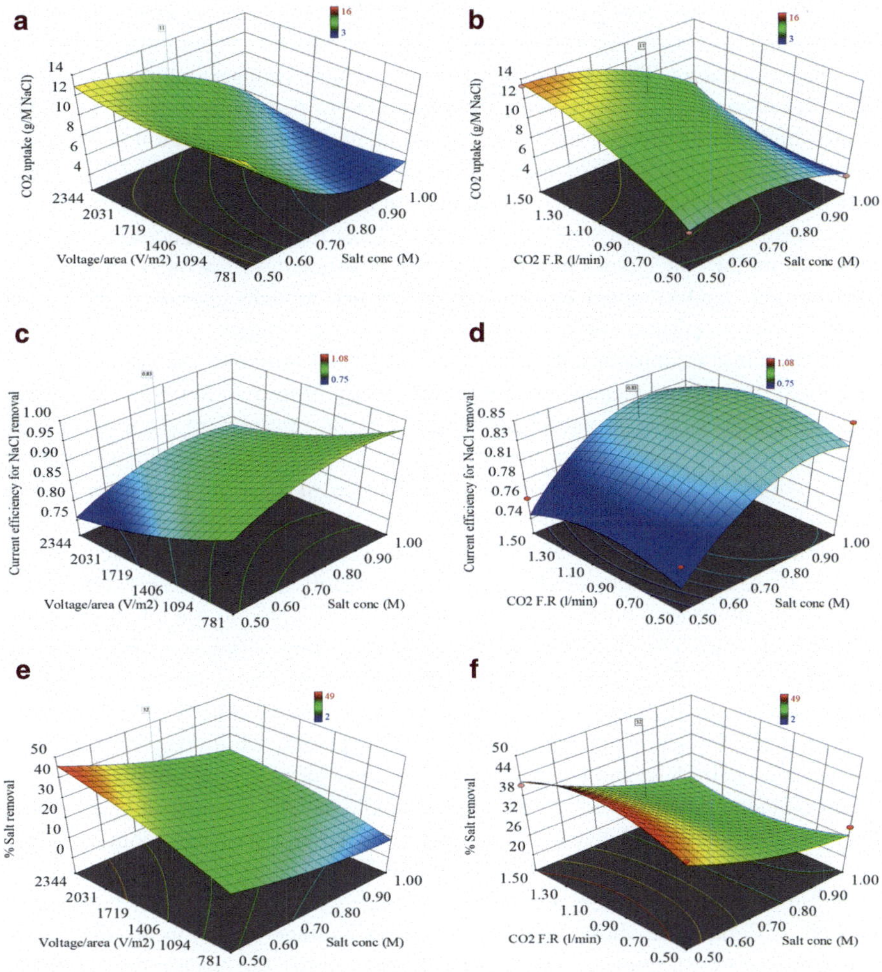

Fig. 16 Effect of parametric variations on CO_2 uptake, current efficiency, and % salt removal in batch circulation configuration. (Mustafa et al. 2023)

0.16 to 1.84 L/min, while keeping the voltage/area at 1562 V/m^2 and NaCl concentration at 0.75 M, the CO_2 uptake experiences an increase from 3 to 10 g/M NaCl. Furthermore, Fig. 16a, b illustrates the impact of NaCl concentration on CO_2 uptake. Optimal CO_2 capture efficiency is attained when the NaCl concentration reaches around 0.75 M, primarily due to the enhanced conductivity in the ED stack. However, exceeding this concentration leads to a decline in CO_2 uptake, potentially due to the occurrence of precipitation. Therefore, it is crucial to maintain high voltage, high CO_2 flow rate, and intermediate NaCl concentrations to achieve a high level of CO_2 uptake.

The impact of different parameters on the current efficiency for NaCl removal is illustrated in Fig. 16c, d. Figure 16c reveals a linear reduction in current efficiency as the voltage increases. This decline in efficiency at high voltages could be attributed to side reactions such as the conversion of H_2SO_4 to peroxydisulfuric acid. Additionally, high voltages can lead to water splitting, an undesired reaction that reduces current efficiency. With an increase in the voltage/area from 248 to 2876 V/m2, while keeping the NaCl concentration at 0.75 M and CO_2 flow rate at 1 L/min, there is a decrease in current efficiency of more than 25%. Figure 16c, d demonstrates the impact of initial NaCl concentration on current efficiency. Elevated initial NaCl concentrations lead to enhanced conductivity, which reduces resistance and consequently improves current efficiency. By raising the NaCl concentration from 0.33 M to 1.17 M, while maintaining the CO_2 flow rate at 1 L/min and the voltage/area at 1562 V/m^2, there is an increase in current efficiency from 0.74 to 0.82. The current efficiency exhibits a linear relationship with NaCl concentration up to approximately 0.83 M, beyond which it starts to decrease, possibly due to precipitation at high concentrations. Figure 16d examines the influence of CO_2 flow rate on current efficiency. With an increase in the CO_2 flow rate, there is a corresponding elevation in the concentration of HCO_3^- and CO_3^{2-} in the gas chamber. This leads to a greater influx of these ions into the alkali chamber. As a result, the osmotic pressure in the salt chamber decreases, enabling a greater passage of Na^+ into the alkali chamber. Consequently, this contributes to an increase in current efficiency. Nevertheless, when the CO_2 flow rate surpasses 0.83 L/min, the current efficiency starts to decline, possibly due to the diffusion of gas into the alkali chamber at high CO_2 flow rates. By raising the CO_2 flow rate from 0.16 to 1.84 L/min, while keeping the NaCl concentration at 0.75 M and the voltage/area at 1562 V/m^2, a slight improvement in current efficiency is observed, increasing from 0.82 to 0.83. In summary, achieving high current efficiency for NaCl removal requires a a high initial NaCl concentration, low voltage, and an intermediate CO_2 flow rate.

The impact of parameters on the percentage of salt removal is depicted in Fig. 16e, f. Figure 16e shows that increasing the voltage results in an increase in the percentage of salt removal. Increased voltages amplify the electric field strength, facilitating the transfer of Na^+ and Cl^- with greater electrical force. Consequently, this promotes the removal of these ions from the salt chamber. By raising the voltage/area from 248 to 2876 V/m^2, while keeping the CO_2 flow rate at 1 L/min and the NaCl concentration at 0.75 M, the percentage of salt removal experiences an increase from 2.3% to 34%. In Fig. 16e, f the decline in salt removal with an increase in the initial NaCl concentration is illustrated. As the NaCl concentration is elevated from 0.33 M to 1.17 M, while maintaining the voltage at 1562 V/m^2 and the CO_2 flow rate at 1 L/min, the salt removal demonstrates a decrease from 49% to 15%. It is important to highlight that the removal of ions increases as the NaCl concentration rises. However, the percentage of salt removal is reduced despite the increase in ion removal. Figure 16f illustrates the effect of CO_2 flow rate on salt removal. As the CO_2 flow rate is raised from 0.16 to 1.84 L/min, while keeping the voltage/area at

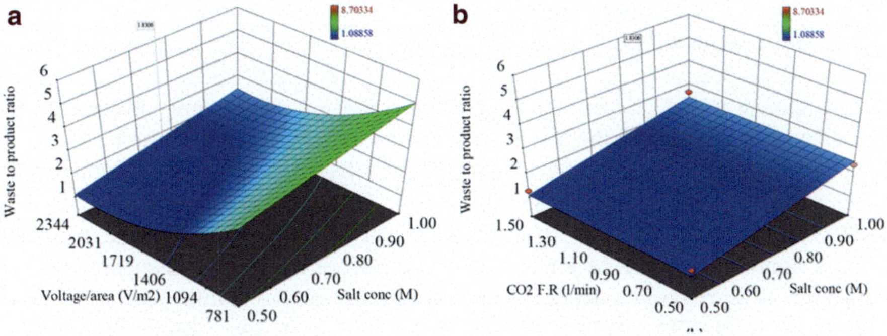

Fig. 17 Effect of parametric variations on waste-to-product ratio in batch circulation configuration. (Mustafa et al. 2023)

1562 V/m^2 and the NaCl concentration at 0.75 M, the overall salt removal experiences a modest increase from 21% to 25%. The removal of salt demonstrates a continuous increase until a CO_2 flow rate of 0.83 L/min is reached. Beyond this point, the salt removal begins to decline. A comparable pattern is noticed regarding the impact of the CO_2 flow rate on current efficiency. In summary, low initial salt concentration, high voltage, and intermediate CO_2 flow rate contribute to enhanced salt removal.

The influence of different parameters on the waste-to-product ratio is depicted in Fig. 17a, b. The waste-to-product ratio demonstrates an increase with higher salt concentrations, suggesting that an escalation in salt concentration does not lead to a proportional rise in the production of desired products. By increasing the salt concentration from 0.33 M to 1.17 M, while keeping the voltage/area at 1562 V/m^2 and the CO_2 flow rate at 1 L/min, the waste-to-product ratio exhibits an increase from 1.09 to 3.53. One possible explanation for this phenomenon is the elevation of osmotic pressure. With an increase in salt concentration, there is a corresponding elevation in the osmotic pressure within the salt chamber. This increase in osmotic pressure results in a decrease in the transfer of Na^+ to the alkali chamber. In contrast, the waste-to-product ratio decreases with higher voltages. By increasing the voltage/area from 248 to 2876 V/m^2, while keeping the CO_2 flow rate at 1 L/min and the NaCl concentration at 0.75 M, there is a decrease in the waste-to-product ratio from 8.7 to 1.8. Elevated voltages promote the transfer of ions with a stronger electric field, leading to an enhanced production of carbonates and HCl. This, in turn, causes a reduction in the waste-to-product ratio. The waste-to-product ratio is not significantly affected by the CO_2 flow rate. Consequently, high voltages and low salt concentrations contribute to lower waste-to-product ratios.

3.4.2 Effects of Parameters on the Product Concentration and Composition

The composition and concentration of the product are significantly influenced by various parameters in both the BMED and batch circulation configurations. Mustafa, Jawad et al. (2021) specifically discussed the effects of such parameters on the product in the batch circulation configuration. The alkali chamber's Na_2CO_3 and $NaHCO_3$ concentrations were determined in each sample using double endpoint titration method with 0.1 M HCl, employing phenolphthalein and methyl orange indicators. With regards to voltage, an increase in voltage resulted in higher concentrations of Na_2CO_3 and HCl. This suggests that at low current density, the transportation of CO_3^{2-} through an AEM with a negatively charged coating is impeded. This hindrance is attributed to the repulsion between the −2 charge of CO_3^{2-} and the negative surface charge of the AEM. Nonetheless, at higher applied voltages, the repulsion between CO_3^{2-} and the negatively charged surface of the AEM can be surpassed by a stronger electrical force. As a result, the transport of CO_3^{2-} is enhanced. Additionally, an increase in voltage enhances the absorption of CO_2 and the transport of HCO_3^- into the alkali chamber, shifting the equilibrium in favor of higher concentrations of CO_3^{2-}. As a result, the formation of Na_2CO_3 intensifies at high voltages, and $NaHCO_3$ is entirely eliminated when the voltage reaches 18.41 V. The increase in voltage also leads to greater formation of H^+ in the anodic cell, resulting in increased production of HCl. Thus, it is preferable to utilize higher voltage in order to achieve high-purity Na_2CO_3 and concentrated HCl. Regarding the effect of NaCl on product distribution, no specific pattern is observed with varying NaCl concentrations. However, lower NaCl concentrations favor Na_2CO_3 formation. The maximum formation of $NaHCO_3$ occurs at an HCl concentration of 0.75 M, while the highest concentration of Na_2CO_3 is obtained at an HCl concentration of 0.33 M. The specific concentrations of each product vary with different NaCl concentrations. Regarding the CO_2 flow rate, the highest formation of $NaHCO_3$ is observed when the flow rate is set at 1.841 L/min. On the other hand, the maximum formation of Na_2CO_3 and HCl occurs when the flow rate is set at 1 L/min.

4 Conclusions and Outlook

The Solvay process is a widely used method for brine management process. However, this process relies heavily on alkali materials that are both environmentally harmful and unsustainable. This dependence on such materials poses a significant challenge in achieving sustainability goals in industrial processes. Additionally, the Solvay process is associated with high costs, which can be attributed to the expense of acquiring alkali materials, regenerating them, and maintaining high operating conditions for CO_2 absorption. In addition to these financial challenges, the by-products of this process have low purity levels, which significantly restricts their potential applications. These issues ultimately undermine the overall efficiency and profitability of the process, making it an unsustainable solution for industrial processes. Thus, there is a growing need to identify alternative processes that are more sustainable and have fewer environmental impacts.

To address these issues and promote sustainable handling of pollutants, an alternative process is necessary for managing reject brine using CO_2. Electrodialysis, an electric field-driven process, has been proposed as a suitable alternative to the conventional Solvay process for reject brine management and CO_2 capture in a combined process. This process utilizes an electric field to separate ions and create high-purity products, including $NaHCO_3$, Na_2CO_3, HCl, irrigation-standard water, and a high-purity NaCl solution from reject brine and CO_2. The high-purity NaCl solution produced is suitable for use in enhanced oil recovery applications. Furthermore, this process can also reduce the water footprint by effectively managing the rejected brine, which conventionally would be disposed of into the open environment, potentially polluting existing water resources.

In BMED, the significant factors included the current density, number of cell pairs, and CO_2 flow rate, and their effect on the process performance was analyzed. The current density had a positive impact on the capture of CO_2 and the percentage of CO_3^{2-} in the carbonate mixture, but it also increased the energy consumption. Increasing the cell pairs reduced the energy consumption, but it also reduced the CO_2 capture. CO_2 flow rate decreased CO_2 capture and increased the percentage of HCO_3^- in the carbonate mixture.

In batch circulation electrodialysis, three process variables (CO_2 flow rate, voltage/area, and salt concentration) were evaluated with respect to their effects on the responses. Overall, CO_2 capture was enhanced by high CO_2 flow rate, high voltage, and intermediate NaCl concentrations. High current efficiency for NaCl removal required high initial NaCl concentration, low voltage, and intermediate CO_2 flow rate. Salt removal was enhanced by low initial NaCl concentration, high voltage, and intermediate CO_2 flow rate. High voltage and low salt concentration resulted in a lower waste-to-product ratio.

The effect of process variables, including current density, flow rate (diluate/concentrate), electrolyte flow rate, number of cell pairs, and membrane type, on the performance of SED was analyzed. Cation-selective membranes in SED were found to be more effective than anion-selective membranes in separation. All current densities showed the highest separation efficiency for Mg^{2+}, followed by Ca^{2+} and SO_4^{2-}. Increasing the current density accelerated the transfer of ions but decreased the separation efficiency and increased energy consumption. A lower flow rate and fewer cell pairs led to greater separation efficiencies for monovalent and divalent ions. Standard commercial membranes were found to consume significantly lesser energy than selective membranes for the separation of ions. The maximum variations were observed for S(Na/Mg), followed by S(Na/Ca) and S(Cl/SO_4).

This chapter presents findings that support the concept of waste-to-resource generation and offers sustainable technology for combined CO_2 capture and reject brine management. This technology enables multi-waste handling with high-purity by-products, providing incentives for industries to consider waste as a renewable resource that can generate profits through the sale or use of by-products as raw materials. The presented technology has the potential to motivate industrialists to adopt sustainable waste management practices and contribute to the development of a circular economy. Moreover, the present project has the potential to address several research gaps by exploring new dimensions, including but not limited to:

- This chapter demonstrates the potential of electrodialysis for multi-waste management and resource generation. The approach presented in the thesis could be applied to other waste streams, opening up new possibilities for sustainable and efficient waste management practices. The thesis presents a novel approach to multi-waste management that involves the use of the electrodialysis process for resource generation. Specifically, the thesis discusses the management of both CO_2 and reject brine through a combined process that results in the formation of value-added products. This innovative approach can be extended further to address other waste streams, such as the management of SO_X and reject brine, which can also yield valuable products. The presence of SO_X in the gas chamber can facilitate the production of sodium sulphite (Na_2SO_3) and sodium hydrogen sulphite ($NaHSO_3$). In addition, the production of high-purity sodium sulphite and sodium hydrogen sulphite can be achieved by fine-tuning the operating conditions. Key factors that can be adjusted to control the selectivity of these compounds include the flow rate of SO_X, current density, and percentage of SO_X in the gas mixture, among others. Sodium sulphite and sodium hydrogen sulphite are versatile chemicals with various applications in industries such as food and beverage, water treatment, photography, pulp and paper, textile, and chemical industries. They are used as preservatives, antioxidants, bleaching agents, reducing agents, and oxygen scavengers, among others. These chemicals are essential in many industrial processes and have unique properties that make them suitable for a wide range of applications. Additionally, the transfer of Cl^- from the anodic cell into the acid chamber can lead to the production of HCl, which can be a valuable product for various industries. The integration of these processes not only facilitates resource recovery but also helps in reducing the overall water footprint associated with industrial waste management.

$$Na_2SO_3 + H_2SO_3 \rightleftharpoons 2NaHSO_3$$

$$NaHSO_3 + H_2O \rightleftharpoons H_2SO_3 + NaHSO_4$$

- The proposed process demonstrated by Mustafa, Jawad et al. for treating reject brine involves two separate stages. The first stage involves SED, followed by either BMED or batch circular electrodialysis. Integrating both stages into one would simplify the process, making it easier to integrate into established industries as a waste-to-resource generation technology. However, this integration would introduce challenges, such as the need for low-resistance selective membranes that can be used in the stack. Typically, commercial selective membranes have high electrical resistance, and the four-chamber design of stage 2 also has high resistance due to multiple chambers. Integrating both stages into one stage and using selective membranes in this stage would further increase the electrical resistance. To reduce electrical resistance, it is necessary to synthesize selective membranes with lower resistance. High resistance increases the osmotic pressure, which can result in the transfer of water from the low-concentration chamber to the high-concentration chamber. To reduce this water transfer, a high current density would be required, increasing the energy consumption of the process. Additionally, the application of a high current density to the IEM can have several negative effects. One of the primary effects is the generation of excessive heat due to resistive heating, which can lead to thermal degradation of the IEM material. Additionally, high current density can cause mechanical stress on the IEM, leading to deformation or even rupture of the membrane. This can result in a loss of selectivity and reduced efficiency of the electrochemical process. Therefore, careful consideration must be given to the design of the process to ensure that the benefits of integration outweigh the potential negative effects.

- The proposed process involves the use of brine, which contains various ions such as Ca^{2+}, Mg^{2+}, Na^{+}, Cl^{-}, K^{+}. However, it is important to note that these reject brine samples used in the proposed process do not contain any organic molecules or hydrocarbons. To increase the versatility of the proposed process, reject brine obtained from various other industries, which may contain organic molecules and hydrocarbons, can also be used. However, the presence of these contaminants can negatively impact the performance of the electrodialysis process in several ways. One of the primary impacts of organic molecules and hydrocarbons on the IEM is fouling and contamination. These contaminants can adhere to the surface of the IEM, leading to a reduction in the ion exchange capacity and overall performance of the membrane. Additionally, the build-up of organic molecules and hydrocarbons can increase the electrical resistance of the membrane, requiring higher energy consumption during the electrochemical process. Furthermore, the presence of these contaminants can cause chemical reactions within the electrochemical process. Oxidation or reduction reactions can occur, leading to the degradation of the IEM material over time. This can result in a decrease in the mechanical strength of the membrane, leading to deformation or even rupture, ultimately resulting in a loss of selectivity and reduced efficiency of the electrochemical process. In addition, organic molecules and hydrocarbons can interact with the functional groups on the surface of the IEM, causing a change in the chemical composition of the membrane. This can alter the ion exchange properties of the membrane, resulting in reduced selectivity and efficiency of the electrochemical process. To prevent these negative impacts on the IEM, it is important to minimize the presence of organic molecules and hydrocarbons in the electrochemical process through proper pretreatment and maintenance of the feed solution, as well as optimizing the operating conditions and parameters during the electrochemical process. Regular monitoring and cleaning of the IEM can also help to prevent fouling and prolong the lifespan of the membrane. Therefore, the future possibility of integrating the proposed process into various industries relies on the proper pretreatment of the reject brine solution to remove organic molecules and hydrocarbons, which can ultimately improve the performance and efficiency of the electrochemical process and reduce water footprint.

- The anion-selective membranes used in the proposed study have demonstrated low separation performance for sulphate ions. As a result, further research is necessary to improve the selective capabilities of anion-selective membranes. One potential approach to enhance the performance of these membranes is to develop new materials with higher selectivity for sulphate ions. Another strategy is to modify the current membrane structure by introducing functional groups or changing the membrane surface charge to enhance its interaction with sulphate ions.
- The proposed process has not investigated the separation of monovalent cations from other monovalent cations, or the separation of monovalent anions from other monovalent anions. This is due to the low concentration of competitive monovalent cations such as potassium in the reject brine samples, as well as the high solubility of potassium-based carbonates/bicarbonates, which prevents the formation of precipitates and avoids interference with the process, unlike calcium and magnesium carbonates/bicarbonates. However, when dealing with reject brine samples that contain high concentrations of other cations and anions with monovalency, the low purity of the resulting by-products and their selectivity would present a challenge when manipulating only the operating conditions. Therefore, additional research is required to address this issue. To ensure successful separation of monovalent cations and anions in high-concentration reject brine samples, it is necessary to investigate new methods and techniques for improving the selectivity and purity of the by-products.
- In the proposed process, multiple chambers have been implemented in stage 2, which has increased the electrical resistance. To reduce this resistance and further improve the efficiency of the process, the use of conducting spacers in electrodialysis can be explored. Conducting spacers could be an alternative to inert spacers, which are currently used and have a high electrical resistance. Conducting spacers can help in reducing electrical resistance in the electrodialysis process, as they provide a more direct path for the flow of electrical current between the electrodes. This can lead to a more efficient process, as less energy is required to overcome the electrical resistance. Moreover, conducting spacers can enhance the mixing and distribution of the feed solution in the electrodialysis stack, resulting in more uniform ion concentration and improved product quality. However, the use of conducting spacers can also introduce challenges such as potential fouling or scaling, which can reduce the effectiveness of the process. Therefore, conducting spacers can be considered as a potential area of future research for the proposed process. By exploring the use of conducting spacers, it may be possible to reduce electrical resistance, improve product quality, and increase the overall efficiency of the electrodialysis process. Further investigation is needed to determine the feasibility and practicality of this approach, as well as to develop appropriate strategies to address any challenges that may arise.
- Ceramic membranes have attracted attention in recent years due to their high chemical and thermal stability, excellent mechanical strength, and superior fouling resistance compared to conventional polymer membranes. Furthermore, ceramic membranes have been found to exhibit lower electrical resistance than

polymer membranes. This property is especially important in electrodialysis, where minimizing electrical resistance is crucial for efficient ion transport and energy consumption. Therefore, testing ceramic membranes with low electrical resistance in the proposed electrodialysis process could lead to improved separation efficiency and lower energy consumption. In addition, ceramic membranes can potentially withstand harsher operating conditions and have a longer lifespan than polymer membranes, making them a promising alternative for electrodialysis applications. However, further research is needed to evaluate the performance of ceramic membranes in electrodialysis and optimize their properties for this specific application.

References

Ahmed M, Shayya WH, Hoey D, Mahendran A, Morris R, Al-Handaly J (2000) Use of evaporation ponds for brine disposal in desalination plants. Desalination. Elsevier.. https://www.sciencedirect.com/science/article/pii/S0011916400000837 Accessed 22 Apr 2024

Ahmed M, Shayya WH, Hoey D, Al-Handaly J (2001) Brine disposal from reverse osmosis desalination plants in Oman and the United Arab Emirates. Desalination 133(2):135–147. Elsevier. https://www.sciencedirect.com/science/article/pii/S0011916401800047. Accessed 22 Apr 2024

Al-Karaghouli A, Renne D, Kazmerski LL (2009) Solar and wind opportunities for water desalination in the Arab regions. Renew Sustain Energy Rev. Elsevier.. https://www.sciencedirect.com/science/article/pii/S1364032108001780?casa_token=d2B3oHtmw-YAAAAA:Au04idcc0-mz2pKiGDiMRD9yafWmQ6jmrrKlA-wS_nvAlDep_hbVYPsqD8QDhxbs_j4FJg7zyA. Accessed 11 May 2024

Al-Kharabsheh S (2006) An innovative reverse osmosis desalination system using hydrostatic pressure. Desalination. Elsevier.. https://www.sciencedirect.com/science/article/pii/S0011916406004346. Accessed 22 Apr 2024

Allen EW (2008) Process water treatment in Canada's oil sands industry: II. A review of emerging technologies. J Environ Eng Sci 7:499–524. https://doi.org/10.1139/S08-020

Al-Sarkal T, Arafat HA (2013) Ultrafiltration versus sedimentation-based pretreatment in Fujairah-1 RO plant: environmental impact study. Desalination. Elsevier.. https://www.sciencedirect.com/science/article/pii/S0011916413000933?casa_token=R3MVECoXWBsAAAAA:K7cOrAZr6LGJNAEg8jN1NVqr0oAZjzy10n_gNdRHh7QLifMy6p2-xYro6tK06XCZQHy6pWBWdw. Accessed 22 Apr 2024

Berreteaga A, Campos E, De Bustamante José I, Iglesias A, Lillo J, Zarzo D (2013) Deep well injection of desalination brines

Burn S, Hoang M, Zarzo D, Olewniak F, Campos E, Bolto B, Barron O (2015) Desalination techniques—a review of the opportunities for desalination in agriculture. Desalination. Elsevier.. https://www.sciencedirect.com/science/article/pii/S0011916415000600?casa_token=Z_8kTSqdPYQAAAAA:cdhnnsTrxi5QNGWdxCh8NjNeyhg3hkVeHSMjbl__6iW7yNMyjiXwSbmZavqcdqHA2CipXd8SiA. Accessed 9 May 2024

Dara S, Lindstrom M, English J, Bonakdarpour A, Wetton B Wilkinson DP (2017) Conversion of saline water and dissolved carbon dioxide into value-added chemicals by electrodialysis. J CO2 Util. Elsevier.. https://www.sciencedirect.com/science/article/pii/S2212982016304620?casa_token=eJm_NLUQ2BUAAAAA:650O5zmTdEDuAgp-w_3hT0kTpRG7tWN3e3o5x1z4VkS1W1ZpPlmhJ17AqKpQM61wafV3dIVGuQ. Accessed 23 Apr 2024

Dara S, Bonakdarpour A, Ho M, Govindarajan R, Wilkinson DP (2019) Conversion of saline waste-water and gaseous carbon dioxide to (bi) carbonate salts, hydrochloric acid and desalinated water for on-site industrial utilization. React Chem Eng. pubs.rsc.org. https://pubs.rsc.org/en/content/articlehtml/2019/re/c8re00259b?casa_token=aIzq9jFJz8oAAAAA:OIT2h_yQduyW2uSkQyGaothA-NbpySPademP833xp7mPkB_GZlzdqkHP2DQK3DSec-P6kg12pQhxdw. Accessed 23 Apr 2024

Elsaid K, Kamil M, Sayed E, Abdelkareem MA, Wilberforce T, Olabi A (2020) Environmental impact of desalination technologies: a review. Sci Total Environ. https://www.sciencedirect.com/science/article/pii/S0048969720350579?casa_token=vlBut40a_TAAAAAA:I7pJkbQcCPHMrwG7xYDA82uFd-Awi0cef0xQiBP_5daitV8r8Yo1ETAUHQkDXgCgoL-9ZbG1CA. Accessed 11 May 2024

Ettouney H, Rizzuti L (2007) Solar desalination: a challenge for sustainable fresh water in the 21st century. Solar Desalination for the 21 St Century. Springer.. https://link.springer.com/chapter/10.1007/978-1-4020-5508-9_1. Accessed 22 Apr 2024

Forstmeier M, Mannerheim F, D'Amato F, Shah M, Liu Y, Baldea M, Stella A (2007) Feasibility study on wind-powered desalination. Desalination.. https://www.sciencedirect.com/science/article/pii/S0011916406012963. Accessed 22 Apr 2024

Ghyselbrecht K, Huygebaert M, Van Der Bruggen B, Ballet R, Meesschaert B, Pinoy L (2013) Desalination of an industrial saline water with conventional and bipolar membrane electrodialysis. Desalination. Elsevier.. https://www.sciencedirect.com/science/article/pii/S0011916413001458?casa_token=sdfv-KpfZ8UAAAAA:3vbd0ucfdyhkMPvHM3-z9x8m-GGEFvN7A-U3bB7VlAEOfL2z9P44JT-WtQf3YyA-BjXk9Ya8vA. Accessed 23 Apr 2024

Gilron J, Folkman Y, Savliev R, Waisman M, Kedem O (2003) WAIV — wind aided intensified evaporation for reduction of desalination brine volume. Desalination 158:205–214. https://doi.org/10.1016/S0011-9164(03)00453-3

Gleick PH, Wolff GH, Cooley H, Palaniappan M, Samulon A, Lee E, Morrison J, Katz D (2006) The world's water 2006–2007: the biennial report on freshwater resources. Google Books.. https://books.google.co.uk/books?hl=en&lr=&id=Lttb1qPh4Z8C&oi=fnd&pg=PR13&ots=cWyhzlE-S0&sig=o1v0WzUaW_88lQGR7QvmjZc3KEg&redir_esc=y#v=onepage&q&f=false. Accessed 22 Apr 2024

Gleick P, Wolff G, Cooley H, Palaniappan M (2013) The world's water 2006–2007: the biennial report on freshwater resources.. https://books.google.com/books?hl=en&lr=&id=Lttb1qPh4Z8C&oi=fnd&pg=PR13&ots=cWyhzlFUY7&sig=0OM4TdQAuEBX6LyUt1kOUR2qkiw. Accessed 22 Apr 2024

Greiter M, Novalin S, Wendland M, Kulbe KD, Fischer J (2004) Electrodialysis versus ion exchange: comparison of the cumulative energy demand by means of two applications. J Membr Sci. Elsevier.. https://www.sciencedirect.com/science/article/pii/S0376738803006148?casa_token=q1aF4PIniIUAAAAA:Wn3zfDNMTIX5BK80f87av4kGW865JJ-wLmFIfugHCQ54x1kL87Vog5rXVT0i2RjpN7HP-rcf0w. Accessed 23 Apr 2024

How Much Water is There on Earth? | U.S. Geological Survey (2019) https://www.usgs.gov/special-topics/water-science-school/science/how-much-water-there-earth?qt-science_center_objects=0#qt-science_center_objects. Accessed 22 Apr 2024

Ihsanullah I, Mustafa J, Zafar AM, Obaid M, Atieh MA, Ghaffour N (2022) Waste to wealth: a critical analysis of resource recovery from desalination brine. Desalination 543:116093. https://doi.org/10.1016/J.DESAL.2022.116093

Iizuka A, Hashimoto K, Nagasawa H, Kumagai K, Yanagisawa Y, Yamasaki A (2012) Carbon dioxide recovery from carbonate solutions using bipolar membrane electrodialysis. Sep Purif Technol. Elsevier.. https://www.sciencedirect.com/science/article/pii/S1383586612004911?casa_token=qphHzvgW8E0AAAAA:8dHILaC1iZZGgXJlkFI0dfhsA9ckywRvYZGs1NETWJQTxVsWkwFtY9K6iGP6Fz-fwCLmvsahZw. Accessed 23 Apr 2024

Jiang C, Wang Y, Wang Q, Feng H, Xu T (2014) Production of lithium hydroxide from lake brines through electro–electrodialysis with bipolar membranes (EEDBM). Ind Eng Chem Res 53(2014):6103–6112. https://doi.org/10.1021/ie404334s

Jones JD Yablonsky A (2016) Carbon dioxide sequestration methods using group 2 silicates and chlor-alkali processes. US9339761 Google Patents.. https://patents.google.com/patent/US9339761/en. Accessed 23 Apr 2024

Jones E, Qadir M, van Vliet MTH, Smakhtin V, Kang S (2019) The state of desalination and brine production: a global outlook. Sci Total Environ. https://www.sciencedirect.com/science/article/pii/S0048969718349167?casa_token=_9FplO7-4CAAAAAA:PDstnXDA11MyL7_zU0lowipzYZD7fxheDdoKpVIJGRTjX_rFs6gBrfj3QITUVdVfkUnZ_lf4EA. Accessed 22 Apr 2024

Kaplan R, Mamrosh D, Salih HH, Dastgheib SA (2017) Assessment of desalination technologies for treatment of a highly saline brine from a potential CO2 storage site. Desalination. Elsevier.. https://www.sciencedirect.com/science/article/pii/S0011916416310086?casa_token=8iYR5Cvy-v8AAAAA:9yRcLXH4rj8tiPtrTMW5W7Wqgr1_GdiCQfqzGy_i0o0isALZ0IyR9Q_-N5_1KDPEQQfSuQA6LA. Accessed 22 Apr 2024

Kelemen PB, Matter J (2008) In situ carbonation of peridotite for CO2 storage. Proc Natl Acad Sci USA 105:17295–17300. https://doi.org/10.1073/PNAS.0805794105

Lee JH, Lee JH, Park IK, Lee CH (2018) Techno-economic and environmental evaluation of CO2 mineralization technology based on bench-scale experiments. J CO_2 Util. Elsevier.. https://www.sciencedirect.com/science/article/pii/S2212982018302403?casa_token=0KaNn4ZxgnsAAAAA:vtlHT6FOhpMABnZ_h8art4QJMWwwMIamyI4bi01daOBGQ2Kf-UAspLchEhLKi1_O4Eqa_DEjwA. Accessed 23 Apr 2024

Lu L, Fang Y, Huang Z, Huang Y, Ren ZJ (2016) Self-sustaining carbon capture and mineralization via electrolytic carbonation of coal fly ash. Chem Eng J. Elsevier.. https://www.sciencedirect.com/science/article/pii/S1385894716310063?casa_token=csLDsyftL3wAAAAA:xYTG_zwha4S8N6wOpGa4dEAQNov1DI0HoXYTQsLW8C07j-sN1cstJmOLfxHUQ3yp-edwdGtV7w. Accessed 23 Apr 2024

Mickley M (2000) Major ion toxicity in membrane concentrate.. https://books.google.com/books?hl=en&lr=&id=Nxluk_xplUgC&oi=fnd&pg=PA1&dq=Mickley,+M.C.+2001.+Major+Ion+Toxicity+in+Membrane+Concentrates.+AWWA+Research+Foundation+Project+%23+290&ots=iOpLD17rPj&sig=cVqEAa8X5iLmhCBpMU0yRHkHWFw. Accessed 22 Apr 2024

Mickley MC (2004) Membrane concentrate disposal:.... Google Scholar. https://scholar.google.com/scholar?hl=en&as_sdt=0%2C5&q=Mickley%2C+M.C.+2004.+Membrane+Concentrate+Disposal%3A+Practices+and+Regulation%2C+Desalination+and+Water+Purification+Research+and+Development+Program+Report+No.+19%2C+U.S.+Department+of+Interior%2C+Bureau+of+Reclamation.+http%3A%2F%2Fwww.mickleyassoc.com%2Fprotected%2Fdisposalfull.htm.&btnG=. Accessed 22 Apr 2024

Mickley M, Concentrate management practices in U.S. municipal desalination plants. World Environmental and Water Resources Congress 2014: Water Without Borders – Proceedings of the 2014 World Environmental and Water Resources Congress (2014) 654–665. https://doi.org/10.1061/9780784413548.070

Mickley & Associates, Environmental Resources Team (U.S.) et al (2018) Membrane concentrate disposal: practices.... Google Scholar.. https://scholar.google.com/scholar_lookup?title=Membrane%20concentrate%20disposal%3A%20practices%20and%20regulation&publication_year=2006&author=M.C.%20Mickley. Accessed 22 Apr 2024

Missimer TM, Maliva RG (2018) Environmental issues in seawater reverse osmosis desalination: Intakes and outfalls. Desalination. Elsevier.. https://www.sciencedirect.com/science/article/pii/S0011916417307750. Accessed 22 Apr 2024

Mustafa J, Mourad A-HI, Al-Marzouqi AH, El-Naas MH (2020) Simultaneous treatment of reject brine and capture of carbon dioxide: a comprehensive review. Desalination 483:114386. https://doi.org/10.1016/j.desal.2020.114386

Mustafa J, Al-Marzouqi AH, El-Naas MH, Ghasem N (2021) Electrodialysis based waste utilization methodology for the desalination industry. Desalination 520:115327. https://doi.org/10.1016/J.DESAL.2021.115327

Mustafa J, Al-Marzouqi AH, Ghasem N, El-Naas MH, Van der Bruggen B (2023) Electrodialysis process for carbon dioxide capture coupled with salinity reduction: a statistical and quantitative investigation. Desalination 548:116263. https://doi.org/10.1016/J.DESAL.2022.116263

Mustafa J, Ghasem N, El-Naas MH, Van der Bruggen B, Al-Marzouqi AH (2024a) Synergistic approach for carbon dioxide capture and reject brine treatment: integrating selective electrodialysis and bipolar membrane electrodialysis. J Clean Prod 438:140578. https://doi.org/10.1016/J.JCLEPRO.2024.140578

Mustafa J, Ghasem N, El-Naas MH, Van der Bruggen B, Al-Marzouqi AH (2024b) CFD modeling for optimal electrodialysis: valorizing reject brine and carbon dioxide for sustainable solutions. Sep Purif Technol 328:125083. https://doi.org/10.1016/J.SEPPUR.2023.125083

Papapetrou M, Epp C Tzen E (2007) Autonomous desalination units based on renewable energy systems-a review of representative installations worldwide. Solar Desalination for the 21st Century. Springer, pp 343–353. https://doi.org/10.1007/978-1-4020-5508-9_25

HS Park, JS Lee, JY Han, S Park, J Park, BR Min 2015, CO2 fixation by membrane separated NaCl electrolysis. Energies. mdpi.Com.. https://www.mdpi.com/1996-1073/8/8/8704. Accessed 23 Apr 2024

Rau GH, Carroll SA, Bourcier WL, Singleton MJ, Smith MM, Aines RD (2013) Direct electrolytic dissolution of silicate minerals for air CO2 mitigation and carbon-negative H2 production. Proc Natl Acad Sci USA 110:10095–10100. https://doi.org/10.1073/PNAS.1222358110

Rhee SW, Reible DD, Constant WD (1993) Stochastic modeling of flow and transport in deep-well injection disposal systems. J Hazard Mater. Elsevier.. https://www.sciencedirect.com/science/article/pii/030438949385097X. Accessed 22 Apr 2024

Saripalli KP, Sharma MM, Bryant SL (2000) Modeling injection well performance during deep-well injection of liquid wastes. J Hydrol (Amst) 227:41–55. https://doi.org/10.1016/S0022-1694(99)00164-X

Shim JG, Lee DW, Lee JH, Kwak NS (2016) Experimental study on capture of carbon dioxide and production of sodium bicarbonate from sodium hydroxide. Environ Eng Res. eeer.Org. https://doi.org/10.4491/eer.2016.042

Skehan S, Kwiatkowski PJ (2000) Concentrate disposal via injection wells–permitting and design considerations. Fla Water Resour J.. https://www.fwrj.com/articles3/0005.pdf. Accessed 22 Apr 2024

Stephens JC, Keith DW (2008) Assessing geochemical carbon management. Clim Change 90:217–242. https://doi.org/10.1007/s10584-008-9440-y

Turek M (2004) Electrodialytic desalination and concentration of coal-mine brine. Desalination. Elsevier. https://www.sciencedirect.com/science/article/pii/S0011916404000694 Accessed 23 Apr 2024

Walha K, Amar RB, Firdaous L, Quéméneur F, Jaouen P (2007) Brackish groundwater treatment by nanofiltration, reverse osmosis and electrodialysis in Tunisia: performance and cost comparison. Desalination. Elsevier.. https://www.sciencedirect.com/science/article/pii/S0011916407000136. Accessed 23 Apr 2024

Xie Z, Ng D, Hoang M, Zhang J, Gray S (2018) Study of hybrid PVA/MA/TEOS pervaporation membrane and evaluation of energy requirement for desalination by pervaporation. Int J Environ Res Public Health 15:1913–1915. https://doi.org/10.3390/IJERPH15091913

Zabolotskii V, Sheldeshov N, Melnikov S (2014) Heterogeneous bipolar membranes and their application in electrodialysis. Desalination. Elsevier.. https://www.sciencedirect.com/science/article/pii/S0011916413005833?casa_token=3OfqBOaeTpsAAAAA:5r8yHYW9vnpbdOuVxXffSDORT4U1Z8LnII4ud0rn6fTAiQKcJEs3474F9DbzLXixd9Qs9f6YgA. Accessed 23 Apr 2024

Index

B
Bipolar membranes, 146, 154, 162, 163, 166–176
Brine management, 5, 6, 145–181
Building life cycle, 118, 126

C
Carbon dioxide capture, 5, 145–181
Clean production, 17

D
Desalination, 5, 6, 146–152, 154

E
Electrodialysis, 6, 146, 152–181
Embodied water, 5, 126, 128
Environmental impacts, 4–6, 12, 13, 17, 19, 20, 23, 26, 29–31, 34, 45, 48–50, 58, 80, 85, 86, 92, 117–119, 121, 125–128, 138, 140, 148, 151, 177
Evaporation, 3, 4, 61, 74, 75, 85–93, 95, 97–102, 104–114, 148, 150, 151

F
Freshwater resources, 1, 2, 17, 57, 58, 146

H
Hydroelectricity, 84, 85, 90–91
Hydropower, 3, 4, 84–86, 90–95, 113

I
Industrial parks, 1, 2, 7–13, 17–19, 29, 35–47, 49
Industrial water use, 39, 45, 46

M
Maintenance, 4, 5, 117–141, 179
Methodologies, 1, 4, 23, 29, 50, 59–61, 63, 85, 92, 93, 95–97, 104, 111, 114, 118, 125–131, 134, 141

S
Selective membranes, 164–166, 177–179
Sustainability, 1–6, 9, 12, 13, 16, 19, 48, 50, 57–80, 84, 85, 95, 113, 118, 122, 123, 125, 128, 176
Sustainable practices, 2, 6, 15, 39, 46

W
Water footprint components, 14, 67, 68, 73–75
Water footprint (WF), 1–51, 57–80, 83–114, 117–141

S. S. Muthu (ed.), *Sustainability and Water Footprint*, Environmental Footprints and Eco-design of Products and Processes, https://doi.org/10.1007/978-3-031-70810-7

Water management, 1, 3, 6, 19, 21–23, 27, 31, 33, 39, 50, 58, 70, 72, 74–76, 79
Water poverty, 2, 3, 58, 59, 62, 70, 78, 79
Water quality, 8, 9, 12–15, 17, 18, 20–23, 26, 29, 31–35, 39–43, 45, 47–50, 147, 151
Water reduction strategies, 59
Water resources management, 63
Water scarcity, 1–3, 15, 29, 57–59, 70, 76, 78–80, 84, 146, 151
Water Stress Index (WSI), 2, 3, 58, 60, 61, 68–69
Wheat and rice, 2, 3, 57–80

The manufacturer's authorised representative in the EU is Springer Nature Customer Service Centre GmbH, Europaplatz 3, 69115 Heidelberg, Germany. If you have any concerns regarding our products, please contact ProductSafety@springernature.com

Printed and bound by CPI Group (UK) Ltd, Croydon, CR0 4YY

01/10/2026

02225980-0014